T0114055

New Tabular Evidence of a Monument in Harmony with the Universe :

A Sourcebook on Nature's Numbers for Artists & Architects

By

Joseph Turbeville

Order this book online at www.trafford.com
or email orders@trafford.com

Most Trafford titles are also available at major online book retailers.

Print information available on the last page.

ISBN: 978-1-4120-1116-7 (sc)
ISBN: 978-1-4122-1653-1 (e)

Trafford rev. 11/29/2019

www.trafford.com

North America & international
toll-free: 1 888 232 4444 (USA & Canada)
fax: 812 355 4082

"The study of the ancient number code is a most delightful pastime; but it is not merely that; nor is it merely of academic antiquarian interest. To investigate number itself in the Pythagorean mode is to become aware of particular numbers which recur constantly in different systems of both numeration and natural phenomena, providing subtle and unexpected links. The emphasis placed on these numbers in all the instruments and products of ancient science, and their enduring reputation as numbers of magical potency, hint that they were formerly regarded as something more than just arithmetical curiosities. It may be that the advanced physicists and cosmologists today, who are beginning to express again the Pythagorean notion that the basic patterns of creation are made up of limited groups of number, will come to realize that they are treading in ancient footprints on paths marked out by long-forgotten predecessors."

"The New View Over Atlantis"
by John Michell

Abstract

The first part of this book will appear familiar to those who might have discovered and read the earlier book *"A Glimmer of Light from the Eye of a Giant"* that was published in the year 2000. It was a small book concerning a revelation that lead to significant archaeological and geological findings.

A simple process of number reduction by digit addition called distillation was applied to the famous Fibonacci number series, and this action produced a limiting effect on the series causing a repetition of the first twenty-four digits, which became the first row of the parent table. This in turn led to the development of several numerical tables whose column and row sums, as well as specific, numerically marked area sums proved in most cases to be historically significant numbers oft times repeated in various other tables.

One of the first discoveries was that several of the tabular sums numerically matched all major external measurements of the Great pyramid of Giza, as measured in feet.

Geological findings in the tables were of things such as Earth and moon size, earth density, rotational speed and rotational energy. These earlier findings and others remain as part of this new edition.

In the second part of this edition, the search for historically significant numbers has been extended to the establishment of a root source for these numbers that seem to have been a part of the "number code" of ancient metrology.

A table of *"products"* developed from the parent table reinforces the Great pyramid's connection to nature's *grand design* by providing a solution of near perfect value for the Golden Ratio Phi(φ), as well as the Royal cubit and its' relation to the number of feet is a statute mile.

The Phi (φ) function wheels provide a source of root numbers who's trigonometric values can be expressed in terms of Phi, as well as all of their rotational values extending to infinity. A rule is developed called *The Trigonometric Phi-function Identity Rule of Natural Symmetry* that identifies those numbers who's trigonometric values can be expressed in terms of Phi(φ). The value of such a rule is presented in the section titled *An Angular Perspective of Leonardo da Vinci Vitruvian Man* (p.71), which delves into the biological angle requirements of the drawing and nature's hidden requirement for such angles.

Nature's numbers are also shown in *Trigonometric Phi Function Rotational Patterns Found in the Wild (*p.77), with a discussion on the five leaf rotational patterns of a springtime weed called Daisy Fleabane.

Over the centuries many great artist and architects have employed the use of the Golden Mean and Golden Rectangle in their creative efforts because of a commonly held belief that by so doing they enhanced the aesthetic value of their work. The same belief might well be assumed to exist by the selection and use of any such number that satisfies the trigonometric Phi-function identity rule.

ii

Contents
PART-1

Glimmer Tables
An Archaeological Find
Without the Aid of Pick and Shovel

PART-2

The Glimmer Tables and the Wheel of Phi (φ).
A Root Source for the Number Code of Ancient Metrology,
From Which the Imperial System of Measure was Founded.

Contents

PART-3

PART-4

List of Tables and Figures

Tables

Figures

Part 1

Part 2

v

PART 1

~~~

The Glimmer Tables

~~~

An Archaeological Find
Without the Aid of Pick and Shovel

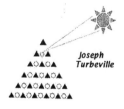

*Joseph
Turbeville*

Preface

If anyone should ask how I managed to conceive the idea and why I decided to develop the tables presented in this treatise, I'm not certain I'll be able to give them a reasonable and acceptable answer. This is an attempt.

I've always considered myself a creative and intuitive person who enjoys the symmetry and beauty of the world around me. A trip to a boatyard or marina to gaze at the sleek hulls of sailing vessels gives me quiet pleasure. The solution to a difficult math problem may sometimes come to me in an intuitive flash, and always seems beautiful. In the last few years I have become increasingly aware of what are called synchronistic events in my life, where coincidental happenings no longer seem merely coincidental but often have real personal meaning or affect the activity I may be involved in. It was in this light that I found myself creating, or quite possibly recreating, these extremely interesting tables that seemed to come to me from out of antiquity and have provided me with a real sense of discovery. Number summations and symmetrical combinations discovered in the tables amazingly represent many of the exact measurements made at the Great Pyramid in Egypt. Other numbers that continually reoccur in the tables represent physical earth measurements of such things as size, density, angular velocity and rotational energy, or harmonics thereof. It seems impossible to imagine that all of these factual numbers were placed here by mere coincidence, when a knowledge of higher mathematics seems essential for their presence, so many so that the reader may be tempted to join in the search for other numerical connections to the distant past. This is my hope and indeed may be the reason for this revelation.

I did not intend this to be a treatise on the synchronistic events that may have led me to this fascinating discovery and these events will only be discussed or mentioned as *Author Side Notes**[*] at those points in this work where I feel it necessary for the reader to be shown how I might have been given a nudge in a fruitful direction by some synchronistic event. This is done in order to minimize any aura of mysticism that might be cast over these tables, for truly they are nothing more than an additive sequence of numbers and combinations thereof that may be found in all of nature.

Table Development

The numerical tables presented in this work were developed by a combination of two well-known mathematical procedures.

- The *Fibonacci series* that appears in many natural growth processes. i.e. 1, 1, 2, 3, 5, 8, 13, 21, 34, 55, ·······
- Numerical reduction of multi-digit numbers by digit addition to produce single digit numbers. This is sometimes referred to as distillation. e.g. 13→4, 21→3, 34→7 or 377→17→8.

The *distilled Fibonacci*[**] numbers and multiples thereof are first used to form nine horizontal rows, each containing 24 digits. Continuation of the rows beyond 24 digits merely reproduces the original digits. Likewise, continuation of the row multiples beyond nine only reproduces each original column.

The tables are shown in differing shades of gray and with black, white and boldfaced numbers for simplifying discussion of table summation processes and to enable the reader to see the patterns of geometrical symmetry.

---

[*]   *Shown in PART-3*
[**]   *Listed in Table-6*

2

Column and row sums are displayed in the white cells along the border of the table. In some cases numbers have been highlighted or enlarged to further aid in viewing the symmetry associated with their location.

In **Table-1** the distilled Fibonacci numbers each occupy a single cell which produces a horizontal cell count of twenty-four. In **Table-2**, the same distilled series digits are placed two to a cell in their original order thereby reducing the horizontal cell count to twelve. These two digits are now treated as double-digit numbers for the column and row summations. A similar row reduction to eight cells occurs for **Table-3** where three series digits occupy each cell and are treated as triple digit numbers when summing.

The tables and sub-tables presented herein display a surprising amount of symmetry, both numerical and graphical, as might be expected by such a modification of this natural mathematical series first reported by Leonardo Fibonacci of Pisa, Italy in the 13th century.

In the development of this treatise, it would have been wrong not to give proper credit to the many great scholars who over the past few centuries have toiled to uncover the mysteries, mathematics and mysticism surrounding the Great pyramid. Exactly who the architects and builders were and when it was built remains a mystery, but whoever they were and whenever it was built, the facts remain, their understanding of what is now known as the Golden ratio Phi(φ), the circumference to diameter ratio of a circle Pi(π), and the dimensions of our planet and our planetary measure of time were all incorporated into the design.

What is now known as Fibonacci's series, even though embedded in the Pyramid's design, was not discovered or found to exist there for centuries after its completion.

If we look at the ratio of successive pairs of numbers from the Fibonacci series, (1, 1, 2, 3, 5, 8, 13, · ·), and divide each number by the one before it, we obtain a series of numbers that approach a limit as the series increases, (1/1=1, 2/1=2, 3/2 =1.5, 5/3 = 1.6⋯, 13/8 = 1.625, 21/13 = 1.61538⋯); ⋯. This limit is called the Golden ratio Phi (φ), or the Golden mean. Its value is ~ 1.618034, and its reciprocal (0.618034) has been shown to be approximately the cosine of the Great pyramid's slope angle. This most interesting number has been incorporated into much of the great art and architecture created in the last few centuries and is said to add aesthetic value by providing a sense of balance and harmony with its presence.

The rhythm of this natural series of numbers can be found in the leaf arrangement on plant stems, various flower petal counts, and the seed head arrangement of sunflowers and pinecones. This is a subject found in the study of botany and is called *phyllotaxis*.

Whatever led me to explore the idea of a numerical reduction of the Fibonacci series and then to the development of these tables probably came from my fascination with numbers and my curiosity of the symmetry, limits, and mirror imagery that I found there.

My first great surprise came when I discovered that **Table-1** contained very visible area and column summations which could provide the major external dimensions of the Great Pyramid as well as very close ratios for Pi $(\pi) \cong$ **594/189 = 22/7** and the Golden ratio Phi $(\varphi) =$ **1.6180** $\cong$ **612/378 = 306/189 = 1.6190.**

The bold lettered, three digit, numbers shown in the above ratios, and many others, will be found to repeatedly occur throughout all tables and area summations in this thesis. Some of the numbers are deemed historically *significant* and have special meanings assigned to them in an ancient science called Gematria. In Jewish mysticism, this is a traditional system of associating numbers with Hebrew letters for discovering hidden meanings in words.

The Great Pyramid - Khufu

The mystery that surrounds the Great Pyramid of Giza in Egypt is legendary, and the differences between truth and legend become less certain as we travel back through the mist of time. It is the most studied and written about monument on the face of the Earth, and it is not my intention to add more verbiage to the vast collection of literature already available on the subject.

It is my intention, however, to show that the tables developed and presented here were inspired by the mathematical symmetry found in nature and for some mysterious reason contain the most precise measurements ever made of the major dimensions of the Great Pyramid *Khufu*. Only after I began to probe the symmetry and mirror image properties of the tables did I begin to wonder; *"Might similar tables have existed before the Great Pyramid was built? Do they contain data or guidelines that might have been used by ancient artisans and architects?"*

Using the width and height measurements recorded by William Petrie[1], an English surveyor of the late nineteenth century as a reference, 440 royal cubits for the base width, 280 royal cubits for the height, and in addition, using his determination of the royal cubit of 20.63 inches, which is now an accepted standard for the Great Pyramid, a base width of 756.4 feet and height of 481.4 feet can be calculated.

There seems to be no direct evidence in the tables of any number that is sufficiently close to the height in feet shown above to be acceptable; however, there are an excess of dimensional numbers from which the height can be calculated. These external numbers can be physically measured on the

---

[1] W.M.Flinders Petrie – The Pyramids and Temples of Gizeh – Chapter 6 –Section 25 – London 1883

pyramid, whereas the height is a number that must be calculated. There is no way to make a direct internal measurement of the height. The base line width **756** feet, the half width **378** feet, the apothem length **612** feet (a measure from the apex down the pyramid face to the half-width point) and the corner-edge line **719** feet, and a close approximation **720** feet, can all be found in the tables.*

- If the readers will examine **Table-1,** they will find four sets of columns in the table that sum to **189.**
- **(4 x 189) = 756.** ⇒ Base width of the Great Pyramid is **756** ft. If the total sum of the numbers of **Table-1 (1188)** is divided by **189,** an ancient irrational value once used for two-Pi $(2\pi_p)$ is obtained. i.e. **1188 / 189** = 44/7 = $2\pi_p$.

- Another way to obtain the pyramid Pi value (π_p), is to subtract the four white corner 9's from the value of either half of the Mirror Image Table-1B to reduce the sum from **414** to **378.** Division of **378** into the total pattern sum of Table-1 **(1188)** produces the pyramid Pi value.

  i.e.   **1188 / 378** = (54 x 22)/ (54 x 7) = **22/7** = π_p

This irrational value of Pi (π_p) is termed the *Pyramid Pi* by the author and is in error from the true value by only 0.04%. It is used throughout this entire treatise because of its tabular presence, and it seems to be the value in use when the pyramid was built.

---

* *All Bold Black numbers can be found in the tables or table summations.*

6

Table-1 Format

Table-1 is a natural formation of **216** Cells arranged in **9** rows of distilled Fibonacci numbers in **24** columns, where each successive row is obtained by multiplying the row number by each respective top row factor and distilling. This limited grid pattern repeats itself when distillation is continued in either the X or Y direction.

Row Sum ▽

1	1	2	3	5	8	4	3	7	1	8	9	8	8	7	6	4	1	5	6	2	8	1	9	117
2	2	4	6	1	7	8	6	5	2	7	9	7	7	5	3	8	2	1	3	4	7	2	9	117
3	3	6	9	6	6	3	9	3	3	6	9	6	6	3	9	3	3	6	9	6	6	3	9	135
4	4	8	3	2	5	7	3	1	4	5	9	5	5	1	6	7	4	2	6	8	5	4	9	117
5	5	1	6	7	4	2	6	8	5	4	9	4	4	8	3	2	5	7	3	1	4	5	9	117
6	6	3	9	3	3	6	9	6	6	3	9	3	3	6	9	6	6	3	9	3	3	6	9	135
7	7	5	3	8	2	1	3	4	7	2	9	2	2	4	6	1	7	8	6	5	2	7	9	117
8	8	7	6	4	1	5	6	2	8	1	9	1	1	2	3	5	8	4	3	7	1	8	9	117
9	9	9	9	9	9	9	9	9	9	9	9	9	9	9	9	9	9	9	9	9	9	9	9	216
45	45	45	54	45	45	45	54	45	45	45	81	45	45	45	54	45	45	45	54	45	45	45	81	1188

189				189				216				189				189				216				

594												594												

Table-1

Table-1 Notes

- Baseline Width (feet) of the Great Pyramid $= 1188 - 2(216) = 4(189) = 756$ feet.
- The total sum minus the sum of all the nines $= 1188 - 9(48) = 1188 - 432 = 756$ feet.
- The digit sum of the two central squares $= 216 \Rightarrow$ Also, 216 is the number of table cells.
- The sum of the perimeter digits (light gray cells) $= 612 \Rightarrow$ **Apothem** length $= 612$ feet.
- The sum of the forty border nines (dark gray on gray cells) $= 9(40) = 360 \Rightarrow 360$ degrees
- Great Pyramid scale size to the Earth $= (1 : 43200) \Rightarrow 2(216) = 432 =$ Base ten factor.
- Pyramid Pi Factor $(\pi_p) = 594 \div 189 = 22/7 = 3.142857$
- Pyramid's Height = Base Area Circumference $\div 2\pi_p = (4 \times 756 \text{ ft}) \div 2(22/7) = 481.1$ feet
- Pyramid's Height = Base Area $\div$ Total Sum of Table-1 $= (756 \text{ ft})^2 \div 1188 \text{ ft} = 481.1$ feet

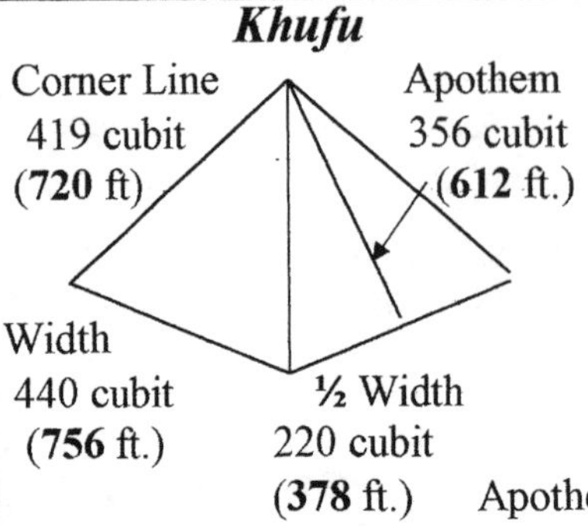

$$\text{Height}^2 = \text{Apothem}^2 - (\tfrac{1}{2}\text{Width})^2$$
$$\text{Height} = \sqrt{((\mathbf{612}\ \text{ft})^2 - (\mathbf{378}\ \text{ft})^2)}$$
$$\text{Height} = 280\ \text{cubit}\ (481\ \text{ft.})$$

The Golden Ratio (φ)

$$\text{Apothem} \div \tfrac{1}{2}\ \text{Width} = 356/220 = 1.618 \cong \varphi$$
$$\text{Phi}\ (\varphi) = (\sqrt{5} + 1)/2 = 1.61803399$$

One cubit = 20.63 inches

Pyramid Pi (π_p)

$$\text{Circumference} \div (2 \times \text{Height}) = (4 \times 440) \div 560 = 22/7 = \pi_p$$

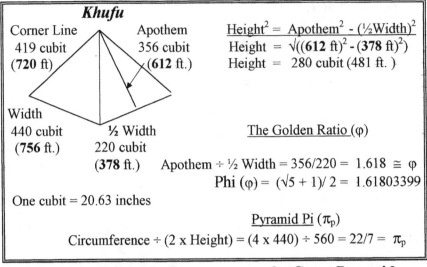

Figure-1. External Measurements of the Great Pyramid. Pi (π_p) and the Golden Ratio Phi (φ).

- By using the apothem **612** ft. and the base half-width **378** ft. from the tables we can determine the slope angle of the Pyramid faces:

$$\text{Cosine of the slope angle } (\beta) = \tfrac{1}{2}\ \text{width} / \text{apothem}$$
$$\text{Cosine } (\beta) = \mathbf{378 / 612} = 0.617647$$
$$\text{Slope angle } (\beta) = 51^\circ.85548690 = 51^\circ\ 51'\ 19.75''$$

- The surface area of a single face of the Great Pyramid is equal to the square of its height:

$$\text{Height}^2 = \tfrac{1}{2}\ (\text{base width}) \times (\text{apothem})$$
$$\text{Height} = \sqrt{((\mathbf{378}\ \text{ft.}) \times (\mathbf{612\ ft}))}$$
$$\text{Height} = 481\ \text{ft.}$$

- The base area of the Pyramid: $(\mathbf{756\ ft.})^2 = 571536$ sq.ft.
  571536 sq.ft. / 43560 sq.ft. per acre = 13.1 acre.

The pattern of numbers displayed in Table-1 is a complete set that will repeat itself in either the X or Y directions if the additions are continued. When this is done, it will be seen that each half of the table (light or dark gray) will be ringed by a band of 42 (9's) whose sum is the base half width of the Great Pyramid measured in feet. This pattern seen in Table-1D is a modified half view of Table-1. The sum of all the white on gray 9's is **378**.

Table-1D   Modified Half View

9	9	9	9	9	9	9	9	9	9	9	9	9
9	1	1	2	3	5	8	4	3	7	1	8	9
9	2	2	4	6	1	7	8	6	5	2	7	9
9	3	3	6	9	6	6	3	9	3	3	6	9
9	4	4	8	3	2	5	7	3	1	4	5	9
9	5	5	1	6	7	4	2	6	8	5	4	9
9	6	6	3	9	3	3	6	9	6	6	3	9
9	7	7	5	3	8	2	1	3	4	7	2	9
9	8	8	7	6	4	1	5	6	2	8	1	9
9	9	9	9	9	9	9	9	9	9	9	9	9

Cell Area	*Cell Count*	*Digit Sum*
• Central Gray Cube	(20 cells)	Digit Sum = **108**
• White Perimeter (Blk. Digits)	(22 cells)	Digit Sum = **99**
• Gray Perimeter (Blk. Digits)	(30 cells)	Digit Sum = **135**
• White 9's in Outer Perimeter	(42 cells)	Digit Sum = **378**
• All Bold Black Digits	(88 cells)	Digit Sum = **414**
Minus 4 Large Corner 9's	– (4 cells)	Digit Sum = – **36**
	(84 cells)	**378**

• Half the baseline width of the Great Pyramid is **378** feet.

Khufu – A Model of the Earth

- The Great Pyramid is considered a scale model representation of the Earth's Northern Hemisphere with a scale factor of (**1:43200**). Its perimeter represents the equator thus the equatorial circumference should be four times the pyramid's base width times the scale factor: i.e.

  4 x **756** ft. x **43200** = 130,636,800 ft.

  130,636,800 ft. ÷ 5280 ft./ s.mi. = 24742 statute miles

  A present day acceptable value is 24904 miles which puts the calculation off by 162 miles. This represents an error of only about one half of one percent, which is a reasonable value.

- Using the height of **481** feet that we calculated in Figure-1 as half of the polar axis of the Earth model and using the assumption that the Earth is a perfect sphere, we can calculate the length of the great circle that passes through the poles: i.e.

  Circumference = $2\pi(481 \text{ ft})(\textbf{43200}) = 130{,}612{,}114$ ft.

  130,612,114 ft. ÷ 5280 ft./mi. = 24737 s. mile

  A present day acceptable value is 24860 statute miles, which puts this calculation off by 123 statute miles. Again, this is an error of about one half of one percent.

- It is pointed out here that on Earth globes and navigational charts imaginary circles are drawn which pass through both geographic poles. These are called Great circles or lines of longitude. On these lines one minute of arc is equal to one nautical mile (~6076 ft.) and there are **21600** minutes of arc in **360** degrees; therefore, the polar circumference of the Earth is **21600** nautical miles, or 24856 s. miles. (Percent error = 0.015)

- The numerical value of the average cell density of Table-1 is found by dividing the total numerical sum (**1188**) by the total number of cells: i.e. **1188 / 216 = 5.5.**
  This number is also the approximate numerical value of the Earth's specific gravity $\cong$ **5.5** i.e. (Density $\cong$ **5.5** gm/cm^3)

- The surface area of the Earth is $4\pi_p R^2$ (mean value), i.e. $(4)(22/7)(3960$ mi. x 5280 ft/mi.$)^2 \cong$ **5.5 x 10^{15}** sq.ft.

- The *normal* to each face of the Great Pyramid is directed exactly toward the four cardinal points of the compass, North, East, South and West respectively. i.e. **360°, 090°, 180°, 270°.**

Notes of Interest Concerning 216

- The number sum of all 40 of the bold white digits in the center rectangles of Table-1 equals **216.**

- The 176 bold black and white digits have a total unit sum **828.** Subtracting the 40 center units (bold white digits) sum of **216** leaves 136 (bold black) border digits with a sum of **612.**
  Note the *mirror image* properties of the numbers **216 & 612.**
  Also, the digit sum of **216** divided by the 40 cell count of the bold white digits gives an average of 5.4 units per cell.
  The remaining boundary unit digit sum of **612** divided by the 136 cell count gives an average of 4.5 units per cell.
  Again, notice the *mirror image* properties with the reversal of the cell densities (4.5 & 5.4).

- The sum of the 24 gray on gray 9's in the 9th row of Table-1 equals **216**.

- The sum of the 20 gray on gray 9's and the 4 central white 9's in each half of Table-1 is **216**.

- The sum of the last four columns in each half of Table-1 is **216**.

- Referring toTables-1B or 1C, observe that there are three separate sets of three adjacent columns which sum to **216**.

- Referring to Table-1C, observe that there are two mirror image rows on each side of the horizontal centerline which sum to **216**.

- The number **216** is not only the total cell count of Table-1 but also, as seen here, seems to manifest itself as the total sum for several different symmetrical summation methods.

- Another note of special interest is the unique fact that a six unit cube is the <u>only</u> single digit cube that has the same numerical value for its surface area as it has for its volume measure. That numerical value is **216**.
  e.g.   6 ft x 6 ft x 6 ft   = **216** ft^3 (volume)
       (36 ft^2/face) x (6 faces/cube)   = **216** ft^2 (surface area)

- Using the only three *Alpha-Numeric* values* from Numerology which distill to six,   F = 6, O = 15, X = 24 as multipliers,   the product **2160** is obtained.
       i.e. 6 x 15 x 24 = **2160**

---

* See Table-5

- The sum of all the interior 90° angles of a rectangular box is **2160°** i.e. 8 corners x 3 angles/corner x 90°/angle = **2160°**

- The Earth's moon has a mean diameter of **2160** statute mile. The Sun's diameter is 400 times the diameter of our moon.
    i.e. 400 x **2160** miles = **864,000** mi.

- The number of seconds in a day is 400 times **216** seconds.
    i.e. 400 x **216** sec = **86400** sec = 1 day

- Two times **216** is numerically equal to the square root of the *speed of light* in miles per second and is in error by ~ 0.2% .
    i.e. **2 x 216 = 432** = $\sqrt{186,624}$ $\Rightarrow$ (186,624 mile/sec)

- The sun is the prime source of light in our planetary system and surprisingly, by using the *Alpha-Numeric* values of the letters in our English word L•I•G•H•T as multipliers and then dividing by their sum we obtain **2160**.
    i.e. $(12 \cdot 9 \cdot 7 \cdot 8 \cdot 20) \div (12+9+7+8+20) = \mathbf{2160}$

- The units that we use today for angular measurement have been brought forward to us from antiquity. They are the degree (°), the arc minute (') and the arc second (").
    One degree (1°) contain (60') arc minutes.
    One arc minute (1') contains (60") arc seconds.

    e.g. **(360 °/ circle)** • **(60 '/ °)** = **21600** '/circle

- On Earth globes and navigational charts, lines of longitude are drawn which pass through both geographic poles. On such lines one minute of arc equals a nautical mile (6076 ft.),
    and the polar circumference equals **21600** n.mile (24856 s.mi.).

An examination of the two halves of Table-1 reveals an inverted duplication pattern for the numbers with the exception of the border nines, hence the choice of bold black and white digits to aid in the discussion. The 40 border **nines** were made gray on gray and were omitted from the Mirror Image diagram of Table-1B and its overlay Table-1C. However, there are four remaining 9's symmetrically centered in each half of Table-1B cornering the two rectangular clusters of 20 cells each. The digits are bold white and each cluster has a sum of **108**. Together their sum is **216**. Notice the summations at the bottom of the table. There are three triple columns that also have a sum of **216.**

The sum of all the digits in Table-1B is **828**. This total sum minus the sum of the white digits (**216**) leaves a remainder of **612,** which is the numerical value of the apothem of the Great Pyramid, measured in feet. Note the numerical mirror image properties of the numbers **216** and **612.**

Table-1C was formed by superimposing the two mirror images of Table-1B directly over one another and summing each of the overlaying **88** cells. The summations at the bottom of the table are naturally identical to those of Table-1B, but what is aesthetically beautiful here is the fact that all of the sums are 9's with the exception of the corner 9's previously discussed which now superimpose as four corner 18's in a single set of 20 dark gray cells. *Truly, there is symmetry in nature!*

The most impressive fact is that the sum of all the 9's happens to be the exact number of feet as the baseline width of the Great Pyramid at Giza, i.e. **756** feet. Again, the sum of all the light gray numbers, now all 9's, is numerically the same as the apothem of the Great Pyramid i.e. **612** feet.

Table-1B

The Right-Half Bold Digits of Table-1 when positioned directly above the Left-Half Bold Digits form a **Mirror Image** pattern that makes the numerical equality of the two sets immediately obvious.

8	8	7	6	4	1	5	6	2	8	1
7	7	5	3	8	2	1	3	4	7	2
6	6	3	9	3	3	6	9	6	6	3
5	5	1	6	7	4	2	6	8	5	4
4	4	8	3	2	5	7	3	1	4	5
3	3	6	9	6	6	3	9	3	3	6
2	2	4	6	1	7	8	6	5	2	7
1	1	2	3	5	8	4	3	7	1	8
1	1	2	3	5	8	4	3	7	1	8
2	2	4	6	1	7	8	6	5	2	7
3	3	6	9	6	6	3	9	3	3	6
4	4	8	3	2	5	7	3	1	4	5
5	5	1	6	7	4	2	6	8	5	4
6	6	3	9	3	3	6	9	6	6	3
7	7	5	3	8	2	1	3	4	7	2
8	8	7	6	4	1	5	6	2	8	1
72	72	72	90	72	72	72	90	72	72	72
216			90	216			90	216		

Total Sum = **828**   Sum of Central Rectangles (White digits) = **216**
Half Sum = **414**   Sum of Perimeters (Bold Black digits)   = **612**

- The apothem length of the Great Pyramid is **612** feet

By superimposing each half of the Mirror Image Table directly over one another and summing the overlaying **88** cells, the following number distribution is obtained that also has mirror image properties.

9	9	9	9	9	9	9	9	9	9	9	198
9	9	9	9	9	9	9	9	9	9	9	
9	9	9	18	9	9	9	18	9	9	9	216
9	9	9	9	9	9	9	9	9	9	9	
9	9	9	9	9	9	9	9	9	9	9	216
9	9	9	18	9	9	9	18	9	9	9	
9	9	9	9	9	9	9	9	9	9	9	198
9	9	9	9	9	9	9	9	9	9	9	
216		90	216		90		216				

©1999, Joseph Turbeville

Dark Gray Area + Light Gray Area = Table Sum

$$16 \times 9 = 144 \qquad 68 \times 9 = 612 \qquad 84 \times 9 = 756$$
$$+ 4 \times 18 = 72 \qquad\qquad\qquad\qquad\qquad + 4 \times 18 = 72$$
$$\underline{216} \quad + \quad \underline{612} \;\; \text{(Mirror Image Numbers)} \quad = \quad 828$$

- The Width of the Great Pyramid at Giza is **756** feet.
- The Apothem of the Great Pyramid is **612** feet.

The difference between these two numbers is **144**, which is the sum of the 16 dark gray 9's in the central rectangle.

At some point in time during the writing of this manuscript, during a quiet period of contemplation, my gaze fell upon a copy of Table-1C in an upside-down position near me in my work area. I realized suddenly that in this position all the nines looked like sixes and the eighteens looked like eighty-ones. Without a moment's hesitation and on a hunch, I added the new numbers, and to my amazement found that their sum was identical to the sum of the original numbers. i.e.

Upright	Inverted
84 x 9 = 756	84 x 6 = 504
4 x 18 = 72	4 x 81 = 324
Sum 828	Sum 828

Continuation of the summing process with the new found numbers, following the same gray scale patterns that revealed Khufu's external dimensions, produces new measurements for a hypothetical pyramid that would be two-thirds the size of Khufu and have the same angle of slope as Khufu, i.e. (51° 51′ 20″).

Table-1 OVL is created when the two halves of Table-1 are overlaid and summed. The results are quite similar to those of Table-1C with the exception that a column and row of eighteens is formed, which further support the argument that the tables have a direct connection with the Great pyramid and its size as an exact scale model of the Earth.

TABLE-1 OVL

When the two halves of Table-1 are placed one over the other and the overlaying digits are added, the following numerical pattern is formed.

Row Digit Sum ▽

9	9	9	9	9	9	9	9	9	9	9	18	117
9	9	9	9	9	9	9	9	9	9	9	18	117
9	9	9	18	9	9	9	18	9	9	9	18	135
9	9	9	9	9	9	9	9	9	9	9	18	117
9	9	9	9	9	9	9	9	9	9	9	18	117
9	9	9	18	9	9	9	18	9	9	9	18	135
9	9	9	9	9	9	9	9	9	9	9	18	117
9	9	9	9	9	9	9	9	9	9	9	18	117
18	18	18	18	18	18	18	18	18	18	18	18	216
90	90	90	108	90	90	90	108	90	90	90	162	**1188**

378	378	432

756

- Base Width of Great Pyramid (Khufu) = **756** feet
- Apothem Length of Khufu = **612** feet = Sum of 68 Light Gray 9's
- True compass heading of Khufu's Northernmost face = **360°**
  Sum of the 20 white on gray border 18's = **360**
- Sum of all 18's = **432** ⇒ Great pyramid to Earth scale (**1 : 43200**)
- Khufu's Pi factor (π_p) = **(1188 / 378) = 22/7 = 3.142857**
- Khufu's Height = $\sqrt{(612)^2 - (378)^2}$ = **481.3** feet

Numerical Comparison to Other Research

A research field that is called Archaeocryptography developed and so named by Carl P. Munck, holds the belief that the pyramids and other ancient monuments around the world are "speaking" to us from the distant past. He has developed a system with which, he claims, we can "decode" the language that they "speak, and thru which, they provide their precise location on his modified Earth grid system. It is Munck's belief that the Prime meridian (0°/360°) passes through the Great Pyramid at Giza rather than Greenwich, England, and his grid system is adjusted for their difference in longitude.

Other writers, Gary Val Tenuta[2], a self-proclaimed "crypto-numerologist" and James Paul Furia , a "New Age" archaeo-cryptographically inclined musician, are members of a group calling themselves "The Code Gang." They are what might be called Munck disciples, and in their own search to discover secrets of the distant past have recovered and employ an ancient numerical science called Gematria that deals with numbers and their correspondence to sound and letters of ancient alphabets.

It is at this point that I would like to propose an addition to Gary Val Tenuta's mystical Grek-5 pyramid that will increase its numerical value and connect it directly to the Great Pyramid at Giza. It involves the digits **1, 8, & 9** that result when he sums and distills each level of his ten level pyramid.

Recalling Table-1, observe that the number **189** appears as the sum of four groups of four columns, and the sum **756** is discussed in the notes as the baseline width in feet of the Great Pyramid. From a historical review by Peter Tompkin[3], when

[2] G. Tenuta – The Secret of Nine – self published 1999 – Code UFO@aol.com
[3] P. Tompkin - Secrets of the Great Pyramid – 1971- Galahad Books publisher – New York, NY

Piazzi Smyth arrived in Egypt in 1865 to study and measure the pyramid, there were great mounds of refuse piled as high as 50 feet around its base. These were the remnants of the destructive removal by the Arabs of its hard limestone casing, which they had used for construction of mosques and palaces in Cairo.

It is very interesting at this point to consider the addition of three more hypothetical tiers to the base of the Grek-5 pyramid (Figure-2), designed to account for "buried tiers". This will produce three more distilled numbers **1, 8**, and **9**, which gives us, (4 x **189**) = **756** = Khufu's width in feet. There would now be 13 tiers on the Grek-5 pyramid, and the base of the Great Pyramid-Khufu is **13** acres. Even more striking is the fact that the number of triangles in the Grek-5 pyramid now totals 169 and the square root of 169 is **13**.

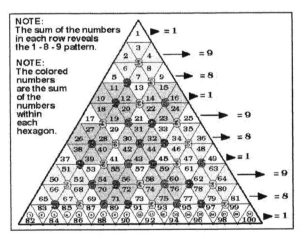

Figure-2. Grek-5 Pyramid Showing the Three Levels of **1, 8, 9.**

□ If the square root of the *Alpha-Numeric* product/sum ratio for the word GREK is calculated, the number 13 appears again.
  i.e. $((7 \times 18 \times 5 \times 11) \div (7+18+5+11))^{1/2} = \textbf{13}$ (See Table-5)

□ In August of 1999, Tenuta learned of a new crop circle formation found near Avebury, England, which seemingly had a striking resemblance to his Grek-5 pyramid. On close examination, the crop circle was found to contain 33 small circles that were arranged in a triangular pattern, each having two or three radial lines so ordered that the lines created the illusion of six raised cubes set down to form their own triangle, each cube showing three faces. Using graphic techniques, Tenuta scaled the image of the glyph so that the radial lines of the small circles were the same size as the hexagon lines of his pyramid, then the six cubes were centered and superimposed atop his triangle forming a nearly perfect match. Really an intriguing discovery, but what does it tell us?

On examination of the overlay by this author, the numbers on the hexagons that lay beneath the three faces of the cubes provided a 3 digit number for each cube when they were added. The sums are:

366 + 390 = 756   **102 + 210 + 234 + 414 = 960**

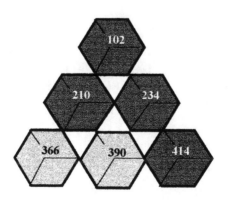

Figure-2A

- The base width of the Great Pyramid is **756** ft.
- Twice the height is ≅ **960** ft., hence the height is ≅ **480** ft.

Because **481** feet is the generally accepted value for the height of the Great pyramid, this answer is in error by only 0.2%. There are probably more comparisons that can be made here, but my intention is only to point out possible similarities in our findings.

❑ On a larger note, the tables provide numerical agreement with some of our solar system's most surprising physical data. For example the sun's diameter is **400** time larger than the moon's and if it were not for this fact, and the fact that the moon's placement relative to the sun causes both objects to appear about the same size to an earth observer, then we could not observe the corona effect around the sun at those rare times of a total solar eclipse. In geometry, both objects are said to share "similar triangles", meaning that the ratio of the moon's radius to its distance from earth is the same as the sun's radius to its distance from earth. i.e.

$$(\textbf{1080} \text{ mi.}/ 236{,}121 \text{mi.}) \cong (\textbf{432{,}000} \text{ mi.}/ 94.5 \times 10^6 \text{mi.})$$

❑ Another interesting fact is that when the Earth's mean diameter (**7920** mi.) is divided by the sum of the Earth and Moon's mean radii (**3960** mi.+ **1080** mi.) = **5040** mi., the half-value of Pyramid Pi (π_p) is obtained. i.e.
7920 / 5040 = 7920 / 7 factorial = 7920 / 7! = $(\pi_p/ 2)$ = 11/7.

❑ If the sum of the Earth and the Moon's mean radii (**5040** mi.) is divided by the Earth's mean radius (**3960** mi.), the value $(4/\pi_p) = \textbf{1.272727}\cdots = \sqrt{\varphi_1}$ is obtained that approximates √Phi.

$\therefore \varphi_1 = (\textbf{1.272727}\cdots)^2 = \textbf{1.619835}$ 0.1% error from true Phi (φ).
A ratio of Table-1 values: $(\textbf{756/594})^2 = \textbf{1.619835} = \varphi_1$

- James Furia[4] in an article discusses a lesser-known fact that the moon has a very unusual period of rotation about its axis that matches its period of revolution about the Earth, (27.3 d) This anomaly causes the Moon to appear stationary (non-rotating) to an Earth observer, i.e. never allowing us to see the "dark side" of the Moon. He wonders about the profound implications of the existence of a mathematically created harmonic order.

- Charles Johnson[5] in an article on measurements of the Great Pyramid, while discussing the Sun's motion relative to the center of the Milky Way in the time frame of the calendrical system employed by the ancient Kemi of Egypt, shows us an extremely interesting calculation that compares the exactness of the fractal results to the baseline measure of the pyramid. i.e. (Number of sec/day)•(Sun's speed relative to Milky Way).

  (86400 sec/day) x (175mi/sec.) = **15120000** mi/day.

  Therefore, the sun travels **7560000** mile in 12 hour, which is a base ten harmonic of the width of the Great Pyramid, **756** ft. From this and other examples he presents, Johnson states "we cannot help but find it extremely suggestive that these exact similarities between distinct events and their measurement may have been by design".

---

[4] *J.Furia – Evidence of a Mathematically Created Solar System –
JamesFuria@aol.com.*
[5] *C. Johnson - Earth/matrix.com - Science in Ancient Artwork, Series No.77. –
The Great Pyramid - p4.*

Earth Measurements and the "Khufu Mile"

In examining a few of the Earth's basic physical properties, some of the Tables' most frequent summations or fractals thereof appear in the results. They are shown here as evidence of their surprising numerical content.

- The tangential speed of rotation of a point on the Earth's equator is easily calculated by multiplying the Earth's angular velocity by its radius.
  $V_t = R \cdot \omega = (3960 \text{ mi})(2\pi \text{ rad/d}) \div (24 \text{ h/d})(60 \text{ min/hr})(60 \text{sec/min})$
  $V_t = \textbf{0.288}$ mi./sec. $= (\textbf{2 x 0.144}$ mi./sec) (See **144** in Table-1C)

- As the Earth rotates on its axis, it has a slight wobble that is referred to as "precession of the equinoxes". It is caused by the gravitational pull of the Sun and Moon on the equatorial bulge and causes the poles to move around a center point called the axis of the ecliptic. The period of a complete rotation is **25920** years.

  In Astrology, the twelve signs of the Zodiac each occupy **30°** of the Astrological circle and encompass a time span of **2160** years. i.e. **12 • 2160** yr. = **25920** year

- The kinetic energy of the Earth's rotation on its axis is $\sim \frac{1}{2}I\omega^2$
  Where: Moment of Inertia (I) is $(2/5) \cdot (M \cdot R^2)$ and the angular Velocity ω is $(2\pi \text{ rad/day} \div (24\text{hr/day})(60\text{min/hr})(60\text{sec/min}))$

  $KE_{rot} = \frac{1}{2}(2/5)(M)(R^2 \cdot \omega^2) = 1/5 \cdot M \cdot V_t^2$
  " $= (1/5)(5.98 \times 10^{24}\text{kg})(214839 \text{ m}^2/\text{sec}^2) = 2.57 \times 10^{29}$ Joule
  " $= (2.57 \times 10^{29}$ Joule$) \div (1.356$ Joule/ft·lb.$)$
  $KE_{rot} \cong \textbf{189} \cdot \textbf{10}^{27}$ ft·lb.   (0.3% error) (See **189** in Table-1)

- The reciprocal of the Earth's angular velocity squared is
  $\omega = (2\pi \text{ rad/day} \div (24\text{hr/day})(60\text{min/hr})(60\text{sec/min}))$
  $(1/\omega^2) = \mathbf{189 \cdot 10^6} \sec^2/\text{rad}^2$.

- In the earlier discussion on *Khufu – A Model of the Earth*, we discussed the fact that 360° had historically been known to contain 21600 arc minutes and that a minute of arc on the Earth's surface was equal to one nautical mile. If we assume that the total sum of Table-2 **(6048)**[*] is Khufu's *inherent* value for the number of feet in a nautical mile, and use that number, which we will now call a ***Khufu mile***, to calculate the distance around the Earth at the equator, we obtain a distance of precisely **21600** *Khufu miles*. i.e.
  The Earth's equatorial circumference is equal to Khufu's perimeter (in feet), times the scale factor, divided by the number of feet per Khufu mile.
  (4 • 756 ft) • (43200) $\div$ **(6048** ft / kfu.mile) $\equiv$ **21600** kfu.mile.

- An interesting problem to consider while discussing the "Khufu mile" is the circumference of a circle with a diameter of **100** kfu.miles. Then express the answer in both feet and statute miles. i.e.
  Circumference = Pyramid Pi(π_p)• Diameter.
  Circumference = (22/7)•(100 ~~kfu.mi~~)•(**6048** ft./~~kfu.mi.~~)
  Circumference = **1,900,800**[▲] ft. = **360** statute mile

  Hence: One degree of arc on this circle is equal to one statute mile, and the radius in statute miles is numerically equal to one-radian.   i.e. **50** kfu.miles = **57.27** statute miles.

---

[*] *John Michell[6] in his writing on ancient metrology provides a list of units that were given two sets of values that arise from the oblate shape of the earth. The nautical mile has the value of **6048** ft. as a minute of latitude at **10°** latitude and **6082.56** ft. as a minute of latitude at **50°**.*

[▲] *See Author Note-2 in Part-3 concerning synchronistic event about 1,900,800.*

□ Corollary: If the diameter of an orb or planet is expressed in units of the Khufu mile, then the number of statute miles per degree on the surface is exactly equal to (1/100) of the numerical value of the diameter.

e.g. Using the Earth's diameter from Fig.3 as an example,
(6917 kfu.mile)•(1/100) = 69.17 s.mile/deg.
(69.17s.mi./deg)•(360deg)=24901 s.mile = circumference
Radius = cir/2π = (24901 s.mile/(44/7)) = 3961.5 s.mile
(This is an error of 0.033 % from the present day mean value)

Additionally, the number **1,900,800** is interesting because it is precisely the number of seconds in **22** days; and the number of seconds in **22** days when multiplied by **22** is exactly the mean diameter of the Earth measured in feet. i.e.

□ Earth mean diameter ≡ **1,900,800** • **22** = 41,817,600.ft.
41,817,600 ft. ÷ 5280 ft./s.mile = **7920** s.mile
Earth mean radius ≡ **3960** s.mile.   (error 0.003 %)

□ An extremely interesting mathematical fact, and one that supports the adoption of the *Khufu mile* (**6048** ft) as a valid pyramid unit arises when one realizes that the numerical value of **9!** (nine factorial) is precisely the number of feet per degree, in the calculation for the Earth's equatorial circumference based on the perimeter measurement of the Great pyramid. i.e.

Perimeter x scale ratio = Equatorial arc length.
(**4•756** ft.)(**43200**) = (130,636,800 ft.) ⇒ i.e. 362880 ft./deg.
362880 ft./deg. ≡ (**9** factorial ft./deg.) = (**9!**) ft./deg.
9! ft./deg. ÷ **6048** ft./kfu.mi. ≡ **60** kfu.mi./deg. (error 0.6%)

- John Michell [6] shows us that the mean radius of the lintel ring at Stonehenge, measured in feet, is a hundredth part of seven factorial (7!), which is 50.4 feet . The mean diameter is therefore 100.8 feet or a sixtieth part of a Khufu mile. Thus the circumference is 316.8 feet or a hundredths part of six statute miles.

- At sea level and at a normal temperature of 79°F. (26°C.), the speed of sound is 1142 ft/sec., which when converted to Khufu units is 0.189 kfu.mi./sec. i.e.

    1142 ft./sec. ÷ **6048** ft./kfu.mi. = **0.189** kfu.mi./sec.

Planetary Measurements and the "Khufu Mile"

Having introduced what we call the "Khufu mile"(**6048** ft), which numerically is the same as the total sum of Table-2, it is now important that we apply this unit to known planetary measurements and examine the results there.

- Figure-3 lists the diameters of the planets and two of the larger moons. The values presented are in "Khufu" units and have been finitely adjusted so that all can be distilled to nine. The average error is less than two tenths of one percent.

- The most surprising result of this examination is that the numerical value of the **756** foot baseline width of the Great Pyramid is a base ten factor of the of the Sun's diameter, **756,000** kfu.mile and also Jupiter's diameter, **75,600** kfu.mile.

---

[6] *J. Michell –The New View Over Atlantis - p.123 – Thames and Hudson Ltd. London, 1993..*

DIAMETER	% ERROR
SUN **756,000 kfu.mi.** = (4000)(189)	x 1.843430400 km/kfu.mi. = 1,393,633 km. True value 1,391,900 km ⟹ error 0.12 %
MERCURY 2646 kfu.mi. = (14)(189)	x 1.843430400 km/kfu.mi. = 4,878 km. True value 4,866 km. ⟹ error 0.25 %
VENUS 6558 kfu.mi. = (34.7)(189)	x 1.843430400 km/kfu.mi. = 12,090 km. True value 12,106 km. ⟹ error 0.14 %
EARTH 6,917 kfu.mi. = (36.6)(189)	x 1.843430400 km/kfu.mi. = 12,751 km. True value 12,742 km. ⟹ error 0.07%
Moon **1890 kfu.mi.** = (10)(189)	x 1.843430400 km/kfu.mi. = 3484 km. True value 3476 km. ⟹ error 0.23 %
MARS 3667 kfu.mi. = (19.4)(189)	x 1.843430400 km/kfu.mi. = 6760 km. True value 6760 km. ⟹ error zero %
JUPITER **75600 kfu.mi.** = (400)(189)	x 1.843430400 km/kfu.mi. = 139,363 km. True value 139,516 km. ⟹ error 0.11 %
SATURN 63,126 kfu.mi. = (334)(189)	x 1.843430400 km/kfu.mi. = 116,368 km. True value 116,438 km. ⟹ error 0.06 %
Moon-Titan 2589 kfu.mi. = (13.7)(189)	x 1.843430400 km/kfu.mi. = 4773 km. True value 4758 km. ⟹ error 0.32 %
URANUS 25,515 kfu.mi. = (135)(189)	x 1.843430400 km/kfu.mi. = 47035 km . True value 46940 km ⟹ error 0.20 %
NEPTUNE 24570 kfu.mi. = (130)(189)	x 1.843430400 km/kfu.mi. = 45293 km . True value 45432 km. ⟹ error 0.30 %
PLUTO 6183 kfu.mi. ≈ (32.7)(189)	x 1.843430400 km/kfu.mi. = 11398 km . True value 11400 km. ⟹ error 0.02 %

Figure-3

Planets and Major Moon Diameters in "Khufu miles"

By most accounts, it has long been known that the Great Pyramid was intended by its mysterious builders to be a scale model of the Northern Hemisphere. They seemingly, during its design and construction, encoded their advanced knowledge of math and science into the monument for the benefit of the generations to come.

Now, with acceptance of this new unit of measure comes the strong probability that another bit of information has been decoded from Khufu's secrets, further supporting the belief that the designers had the knowledge to precisely measure the universe, as has been shown by the numerical match of the Pyramid's width and the Sun's diameter.

- In the earlier discussion of **216,** it was shown that (2 x **216**) squared was equal to the speed of light expressed in statute miles per second. i.e. 186,624 s.mile/sec. (0.11 % error) Conversion of this number to the Khufu unit gives 162,925.7 and with a finite adjustment of plus 0.4 units, it gains mirror image symmetry around the nine and can also be distilled to nine.
  i.e. Speed of Light = 162,926.1 kfu.mi./sec.

PART 2

~~~~

The Glimmer Tables and the Wheel of Phi (φ), A Root Source for the Number Code of Ancient Metrology from Which the Imperial System of Measure was Founded.

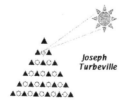

Joseph
Turbeville

Historically Significant Numbers Found When Symmetrically Spaced Column Digits of Table-1 are Cubed and Summed.

Each of four vertically adjacent column digits of Rows 3, 4, 5, & 6 of Table-1[*] that have a sum of 18, produce historically significant numbers when the four digits are individually cubed and summed.

Likewise, the two remaining pairs situated above and below the four digit sets referenced above also produce historical significant numbers when their individual values are cubed and summed. i.e.

- Digits 6, 8, 1, 3 and their common column pairs 2, 4, & 5, 7 located above and below 6, 8, 1, 3, produce the following sums.

$$6^3 + 8^3 + 1^3 + 3^3 = 756 \qquad \& \qquad 2^3 + 4^3 + 5^3 + 7^3 = 540$$

There are four (**4**) columns that produce these two number sets.

$$\therefore \quad 4 \times 756 = 3024 \qquad \& \qquad 4 \times 540 = 2160$$

- Base width of the Great pyramid = **756** ft.
- Perimeter of the Great pyramid = **4 x 756** feet = **3024** ft.
- Moon Diameter = **4 x 540** s.mi = **2160** s.mi = **11,404,800** ft
  Moon Diameter = π_p x **10!** ft. = **6 x 1,900,800** ft.

- Digits 3, 4, 5, 6, and their common column pairs 1, 2, & 7, 8, located above and below 3, 4, 5, 6 produce the following sums.

$$3^3 + 4^3 + 5^3 + 6^3 = 432 \qquad \& \qquad 1^3 + 2^3 + 7^3 + 8^3 = 864$$

There are ten (**10**) columns that produce these two number sets

$$\therefore \quad 10 \times 432 = 4320 \qquad \& \qquad 10 \times 864 = 8640$$

Sum = **12960** yr = Half Earth's ecliptic cycle of **25920** yr

---

[*] Gray scale adjusted Table-1 follows

Table-1   (Gray Scale Modified for this discussion)

216 Cells arranged in **9** rows of distilled Fibonacci numbers in **24** columns, where each successive row is obtained by multiplying the row number by each respective top row factor and distilling.

(This is a limited grid pattern that repeats itself if continued in either the X or Y direction.)

ROW SUM ▽

1	1	2	3	5	8	4	3	7	1	8	9	8	8	7	6	4	1	5	6	2	8	1	9	117
2	2	4	6	1	7	8	6	5	2	7	9	7	7	5	3	8	2	1	3	4	7	2	9	117
3	3	6	9	6	6	3	9	3	3	6	9	6	6	3	9	3	3	6	9	6	6	3	9	135
4	4	8	3	2	5	7	3	1	4	5	9	5	5	1	6	7	4	2	6	8	5	4	9	117
5	5	1	6	7	4	2	6	8	5	4	9	4	4	8	3	2	5	7	3	1	4	5	9	117
6	6	3	9	3	3	6	9	6	6	3	9	3	3	6	9	6	6	3	9	3	3	6	9	135
7	7	5	3	8	2	1	3	4	7	2	9	2	2	4	6	1	7	8	6	5	2	7	9	117
8	8	7	6	4	1	5	6	2	8	1	9	1	1	2	3	5	8	4	3	7	1	8	9	117
9	9	9	9	9	9	9	9	9	9	9	9	9	9	9	9	9	9	9	9	9	9	9	9	216
45	45	45	54	45	45	45	54	45	45	45	81	45	45	45	54	45	45	45	54	45	45	45	81	1188

189	189	216	189	189	216
594			594		

TABLE-1

Table-1 Notes

- Baseline Width (feet) of the Great Pyramid = **1188 – 2(216)** = **4 (189)** = **756** ft.
- The total sum minus the sum of all the nines = **1188 - 9(48)** = **1188 – 432** = **756** ft.
- The digit sum of the two central squares = **216** ⇒ Also, **216** is the # of table cells.
- The sum of the perimeter digits (light gray cells) = **612** ⇒ **Apothem** length = **612** ft.
- The sum of the forty border nines (dark gray cells) = **9(40)** = **360** ⇒ **360** degrees
- Great Pyramid scale size to the Earth = **1: 43200** ⇒ **2(216)** = **432** = Base ten factor.

- Pyramid Pi Factor (π_p) = **594 ÷ 189** = **22/7** = **3.142857**

- Pyramid Height = Base Area Circumference ÷ $2\pi_p$ = (4 x **756** ft) ÷ 2(22/7) = **481.1** ft.
- Pyramid Height = Base Area ÷ Total Sum of Table-1 = (**756** ft)^2 ÷ **1188** ft = **481.1** ft.

- Digits 6, 2, 7, 3 and their common column digits 5, 1, & 8, 4 located above and below 6, 2, 7, 3 produce the following sums.

  $6^3 + 2^3 + 7^3 + 3^3 = 594$ & $5^3 + 1^3 + 8^3 + 4^3 = 702$

  There are four columns that produce these two number sets.

  $$4 \times 594 = 2376 \quad \& \quad 4 \times 702 = 2808$$

- **2376** ft. = Circum. of the largest inner circle of the GP base area.
  Dia. = Circum. / π_p = **2376** ft. / **(22/7)** = GP base width = **756** ft.
  2376 / 2 = **1188** = Total digit sum of parent Table-1.

- Golden Ratio (φ) $\cong$ (GP perimeter $\div$ GP inner circle circum.)^2
  (**3024** ft. / **2376** ft.)^2 = (1.272727..)^2 = 1.619835 = **Phi (φ_1)**
  True Value **Phi (φ_T)** = 1.618034 $\therefore$ Phi (φ_1) error ~ 0.11%.

- A unique *Numerical Mirror Image pair*: **59.4** inch = **4.95** ft.

- It should be noted that the sum of the cubed digits is the same for all three sets. i.e.

  $$(756 + 540) + (432 + 864) + (594 + 702) = 3 \times 36^2 = 3888$$

  Nineveh Number* / **3888** $\equiv$ **7 factorial x 10^7 seconds**

  1.959552×10^{14} sec./ **3888** $\equiv 7! \times 10^7$ seconds
       "   " $\equiv$ 14 million hours
       "   " $\cong$ 1600 years (365 d/y)

---

* M. Chatelain – "Our Cosmic Ancestors" – ISBN 0-929686-00-4 -- p.34 -- Shows that this number discovered on a tablet, recovered from an archeological dig near Nineveh, the ancient capital of Assyria , represents an astronomical span of time approximately 240 times the Earths ecliptic cycle (~25920 years).

Table of Products Reinforces the Great Pyramid
Connection to Nature's Grand Design

In a continuing effort to identify historically significant numbers, and to unravel the mathematical patterns of nature that provide evidence for the existence of a "grand design", a table of products is formed here from the same distilled digits of the Fibonacci series that were used to form Table-1.

If each distilled row digit of Table-1 is multiplied by the digit that follows it, and the product is then distilled and used in place of the multiplied digit, a new and unique table is formed. This new table (Table-7) follows, and is limited to nine rows of twenty-four cells per row, just as its parent table. Portions of the table have been shaded a darker gray than other sections to aid in locating the numbers and digits under discussion.

It should be noted that the first eight rows have a horizontal dividing line separating a four-row mirror-image pattern. A vertical dividing line separates two identical panels, each with **108** cells.

- The first three columns of these two identical panels contain three side-by-side digits that when added sum to **612**. This sum in feet is an apparent major design dimension of the Great pyramid. i.e. **126 + 486 = 612**

612 feet was earlier assumed the intended design length of the apothem[*] because of the geometrical pattern sums found in the parent table. The number is again strongly suggested by the findings in Table-7, as there are four sums of the number **612** that might suggest correspondence to the four faces of the Great pyramid.

---

[*] The apothem is the measure from the apex down the pyramid face to the mid-point of the baseline.

Table-7 Format

216 Cells formed with 9 Rows of the distilled products of the original 24 distilled Fibonacci numbers of Table-1. Each successive row is obtained by multiplying the square of the row number by each respective top row factor and distilling. Some cells in the table have been darkened to aid in the discussion of historically significant numbers discovered here.

TABLE-7

ROW SUM ▽

																								ROW SUM
1	2	6	6	4	5	3	3	7	8	9	9	1	2	6	6	4	5	3	3	7	8	9	9	126
4	8	6	6	7	2	3	3	1	5	9	9	4	8	6	6	7	2	3	3	1	5	9	9	126
9	9	9	9	9	9	9	9	9	9	9	9	9	9	9	9	9	9	9	9	9	9	9	9	216
7	5	6	6	1	8	3	3	4	2	9	9	7	5	6	6	1	8	3	3	4	2	9	9	126
7	5	6	6	1	8	3	3	4	2	9	9	7	5	6	6	1	8	3	3	4	2	9	9	126
9	9	9	9	9	9	9	9	9	9	9	9	9	9	9	9	9	9	9	9	9	9	9	9	216
4	8	6	6	7	2	3	3	1	5	9	9	4	8	6	6	7	2	3	3	1	5	9	9	126
1	2	6	6	4	5	3	3	7	8	9	9	1	2	6	6	4	5	3	3	7	8	9	9	126
9	9	9	9	9	9	9	9	9	9	9	9	9	9	9	9	9	9	9	9	9	9	9	9	216

234			198			270			234			198			270			1404
702									702									

DISTILLED ROW-DIGIT PRODUCTS OF PARENT TABLE-1

TABLE-7 NOTES

- Four dark gray "Tees" have the GP corner-line length of **719** ft. & base half-width of **378** ft. ∴ Apothem = $\sqrt{(719^2 - 378^2)}$ = 611.6 ft. ≅ **612** ft
- Also, sum of **126 + 486 = 612** ⇒ **612** ft. = Apothem of GP.
- Six rows sum of **126 = 756** ⇒ GP's base width is **756** ft.
- Two rows sum of **216 = 432** ⇒ Earth to GP scale ratio **43200 : 1**
- The fourth and fifth row of each panel starts with the number **756**
- The baseline width of the Great pyramid is **756** feet.

- As part of the gray shaded "tee" in the mirror portion of the table, the number **719** makes up the vertical portion of the 'tee' and the number **378** the crossbar; the *seven* digit being common to both numbers. A corner line of the Great pyramid is **719** feet, and the half-baseline width is **378** feet. Using the theorem of Pythagoras, another measure of the Apothem length is calculated. i.e

Apothem length $= \sqrt{(719^2 - 378^2)}$ ft^2 = 611.618345 ft. $\cong$ **612** feet.

- The ratio of this exact calculated apothem value to the tabular half-base width provides a near perfect value for the Golden Ratio.
  Phi (φ) = 611.6183 ft. / **378 ft.** = **1.618038**   (0.00025% error)

- Using the theorem of Pythagoras with the tabular apothem length and the half-width of the baseline, the height of the Great pyramid is calculated. i.e.

Height $= \sqrt{(612^2 - 378^2)}$ft^2 = **481** feet.

- The surface area of one face of the pyramid is equal to the half-base width multiplied by the apothem. i.e.
  Surface Area = **378** ft. x **612** ft. = **231,336** ft^2.

- The square root of the surface area of one face of the pyramid is also equal to the design height. i.e.

Height $= \sqrt{(231{,}336 \text{ ft}^2)}$ = **481** feet

It is amazing that these five three-digit numbers, **126, 486, 756, 378** and **719**, form symmetric patterns and produce Pythagorean solutions that are major dimensions of the Great pyramid.

Could the designers have developed similar tables from nature, and selected the dimensions for the pyramid from the numerical patterns and symmetry they found there? Surely, they must have believed that future generations would recognize the simplicity of their mathematical planning and acknowledge the great influence that nature had on the pyramid design.

A significant finding, buried in the mirror image portion of this table, is the accepted value of the Royal cubit, and it is within the range recorded by W.M. Flinders Petrie in his historic work, *The Pyramids and Temples of Gizeh.*[1] i.e.

Royal Cubit = **20.620 ± 0.005** inches (British Imperial Units).

An examination of Table-7 reveals the four panels of forty-four cells whose digits form two identical Mirror-Image sections. The border nines of columns 12 & 24 and the bottom row of nines are not considered as part of the mirror image panels.

. When the squares of the digits in the eleven cells of either section of the twelve mirror image Rows 1 & 8, 2 & 7, and 4 & 5 are added, it will be found that each set of eleven cells has the sum of 330 units. When multiplied by twelve, the numerical value of the Earth's mean radius, as measured in statute miles, is obtained.

cell-11

$12 \sum_{\text{cell-1}} n^2 = 3960 \Rightarrow$ **3960 s.miles = Earth's Mean Radius**

---

[1] The Pyramids and Temples of Gizeh - Chapter 20- Section 136 - Published London 1883

- When the digits in the eleven cells of either section of mirror image Rows 1 & 8, 2 & 7, or 4 & 5 are multiplied by four, then squared and summed, the numerical value is the number of feet in a statute mile.

cell-11
$$\Sigma_{cell-1}(4n)^2 = 5280 \quad \Rightarrow \quad \textbf{5280 feet = Statute Mile}$$

- Dividing the cell digits in the eleven cells of either section of mirror image Rows 1 & 8, 2 & 7, or 4 & 5 by a factor of four and then squaring, leads to the numerical value of the Royal cubit in inches.

cell-11
$$\Sigma_{cell-1}(n/4)^2 = 20.625 \quad \Rightarrow \quad \textbf{20.625 inches = Royal Cubit}$$

Table-7 thus provides a value for the Royal cubit that is in the range recorded by W.M. Flinders Petrie, **20.620 ± 0.005** inches, and a strikingly simple numerical connection between the Royal cubit in inches and the statute mile in feet.

$$\text{i.e.} \quad (16)^2 \text{ x } 20.625 = 5280$$

It has been pointed out that the value *756* feet was assumed the design base-width of the Great pyramid because of its significant presence in these tables. When expressed in inches and divided by 440 Royal Cubit, it provides a value for the Royal Cubit of $20.618\cdots$. This value is also within the range selected as the best value by W.M. Flinders Petrie.

The most amazing finding in Table-7 concerns the two rows that start with the digits **7-5-6**. Now, note that the next four digits in these two rows, when cubed and summed have a value of *756*. Adjacent to those digits, the last four digits in this mirror image portion of the table, when cubed and summed, have the value *828*. i.e.

$6^3 + 1^3 + 8^3 + 3^3 = 756$ & $3^3 + 4^3 + 2^3 + 9^3 = 828$

Note that the number **828** is the digit sum of the mirror image portion of Table-1, the genesis of all the tables.

In what was seemingly a synchronistic event, the number *756* was found eight times in these two rows of Table-7. This raised the thought that something of historical significance might be found here. It was quickly discovered that the sequence of digits **6-1-8-3** adjacent to *756*, if assumed a four-digit number in statute miles, was close to an equatorial quadrant of the Earth. Multiplying **6183** s.miles by four and applying the Earth: Pyramid scale factor provides a measure for the baseline circumference and width of the Great pyramid that is extremely precise. i.e.

Equator = **4 x 6183** s.mile = **24732** s. mile = **130,584,960** feet.

Great pyramid circum. = **130,584,960** feet/ **43200** = **3022.8** feet
The Great Pyramid width = **755.7** feet = **9068.4** inches
Mean width recorded by W.M.F. Petrie [*] = **9068.8 ± 0.5** inches

It should be remembered that the dimensions extracted from the tables developed here are not measurements made at the Great pyramid, but their finding surely provides mathematical support for the *intelligent design movement* of creationists, and supports the argument for the existence of a "universal grand design". This same value for the equatorial circumference can also be found in the six rows sums of Table-3D that sum to 4122. i.e.

6 x 4122 s.mile = **24732** s. mile = **130,584,960** feet.

---

[*] The Pyramids and Temples of Gizeh - Chapter 6-Sect. 25 - Published London 1883

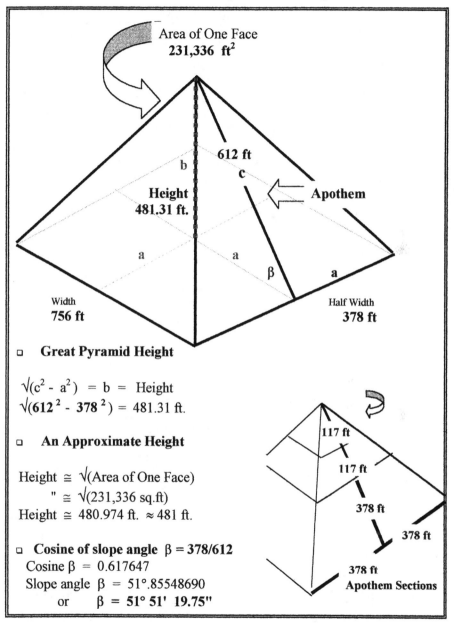

Area of One Face
231,336 ft^2

612 ft

b

c

Height
481.31 ft.

Apothem

a   a

β   a

Width
756 ft

Half Width
378 ft

□ **Great Pyramid Height**

$\sqrt{(c^2 - a^2)} = b = \text{Height}$

$\sqrt{(612^2 - 378^2)} = 481.31 \text{ ft.}$

□ **An Approximate Height**

$\text{Height} \cong \sqrt{\text{(Area of One Face)}}$

$" \cong \sqrt{(231,336 \text{ sq.ft})}$

$\text{Height} \cong 480.974 \text{ ft.} \approx 481 \text{ ft.}$

□ **Cosine of slope angle** $\beta = 378/612$

$\text{Cosine } \beta = 0.617647$

$\text{Slope angle } \beta = 51°.85548690$

or $\beta = 51° 51' 19.75"$

117 ft

117 ft

378 ft

378 ft

378 ft
Apothem Sections

Figure-4   Great Pyramid's Tabular Surface Area Geometry

43

Utilizing tabular apothem sectional data shown in the insert of Figure-4 an extremely accurate value of Phi (φ) can be determined. i.e.

$\sqrt{(612 / (117 + 117))}$ = 1.617215080 $\cong$ Phi (φ), but smaller.
(0.05 % error)

612 / 378 = 1.619047619 $\cong$ Phi (φ), but larger.
(0.06 % error)

If these two conjoined measurements are averaged, an extremely accurate value for Phi (φ) is obtained.

Average value = (1.617215080 + 1.619047619) / 2 = 1.618131350
True value = 1.618033989
Diff. = 0.000097361

0.006 % error

Geometrical Comparison:

Let X^2 = **612 / (117 + 117)** = **612 / 234** = 2.615384615 $\cong$ φ^2
Let Y = **378 / (117 + 117)** = **378 / 234** = 1.615384615 $\cong$ φ
Diff = 1.000000000

For Phi (φ) Golden Ratio

$\varphi^2 - \varphi - 1 = 0$
$\varphi^2 = \varphi + 1$
$\varphi = \sqrt{(\varphi + 1)}$
$1 = (\varphi - 1/\varphi)$

For the X & Y values above:

$X^2 - Y - 1 = 0$
$X^2 = Y + 1$
$X = \sqrt{(Y + 1)}$
$1 = (X^2/Y - 1/Y)$

φ = 1.618033989   X = 1.617215080   1/X = 0.618346942
$1/\varphi$ = 0.618033989   Y = 1.615384615   1/Y = 0.619047619

The Missing Capstone of the Great Pyramid

The most certain, possibly original, design height and baseline slope angle of the Great pyramid was **481.3** feet with a slope angle of **51° 51' 19.75''**. These were from calculations using the tabular baseline half-width of **378** feet and apothem length of **612** feet. The calculations are shown as part of Figure-4.

An examination of the measurements recorded by W.M.Flanders Petrie *(circa)* 1883, atop the Great pyramid at the 203rd course level, provides this truly remarkable comparison, which lends credence to the value of the these tables. It should be noted that course levels 202 and 203 were never completed as is evidenced by Petrie's measurements.

The difference between our tabular based original design height of 481.3 feet and Petrie's height determination at the 203 course, would therefore be the height of the final missing section. i.e.

Tabular based design height = 481.3 ft   = 5,775.6 inch
Petrie's measure at the 203 course (NE level) = 5,451.8 inch
Missing capstone(s) height ⇒   =   323.8 inch
323.8 inch = 26.98 ft ≅ **27** ft.

An examination of the cross sectional view of the Great pyramid displayed in Figure-4A reveals one possible geometrical configuration using summary data from Table-1.

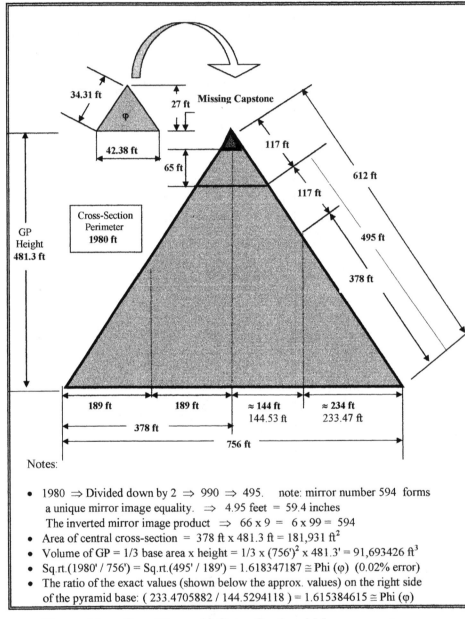

Notes:

- 1980 ⇒ Divided down by 2 ⇒ 990 ⇒ 495.   note: mirror number 594 forms
  a unique mirror image equality.   ⇒ 4.95 feet = 59.4 inches
  The inverted mirror image product ⇒ 66 x 9 = 6 x 99 = 594
- Area of central cross-section = 378 ft x 481.3 ft = 181,931 ft^2
- Volume of GP = 1/3 base area x height = 1/3 x (756')^2 x 481.3' = 91,693426 ft^3
- Sq.rt.(1980' / 756') = Sq.rt.(495' / 189') = 1.618347187 ≅ Phi (φ) (0.02% error)
- The ratio of the exact values (shown below the approx. values) on the right side
  of the pyramid base: (233.4705882 / 144.5294118) = 1.615384615 ≅ Phi (φ)

Figure-4A   **Great Pyramid Cross-Sectional Measurements**
Showing Historically Significant Numbers

46

A diagram of the Great pyramid's base area geometry, including the perimeter and area of the inner inscribed circle, reveals the number **1188** that is the sum total of Table-1. The number **1188**, in units of feet, represents the half-circumference of the inner circle.

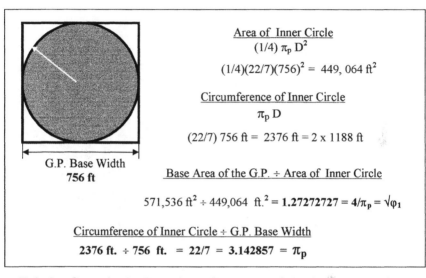

Area of Inner Circle
$(1/4)\,\pi_p\,D^2$

$(1/4)(22/7)(756)^2 = 449,064\ ft^2$

Circumference of Inner Circle
$\pi_p\,D$

$(22/7)\ 756\ ft = 2376\ ft = 2 \times 1188\ ft$

Base Area of the G.P. ÷ Area of Inner Circle

$571,536\ ft^2 \div 449,064\ ft.^2 = \mathbf{1.27272727} = \mathbf{4/\pi_p} = \sqrt{\varphi_1}$

Circumference of Inner Circle ÷ G.P. Base Width
2376 ft. ÷ 756 ft. = 22/7 = 3.142857 $= \pi_p$

G.P. Base Width
756 ft

Tabular Sums in the Base Area Geometry of the Great Pyramid
Figure-5

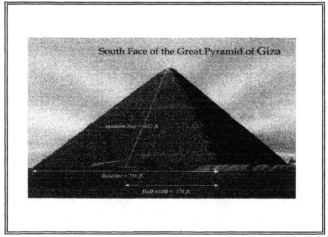

The Phi (φ) Function Wheel - A Cosmological[1] Connection

On examination of Figure-6, the reader will observe a twenty-spoke wheel.   Each spoke displays three angular values of (**n**) that represent repetitive wheel rotations.   Sixteen spokes have angular values that are expressed as trigonometric functions in terms of Phi (φ).   The remaining four spokes represent the cardinal points of a compass, (90°, 180°, 270° and 360°), and have a Sine or Cosine value of either zero or ± 1.

For clarity, and to separate the Sine and Cosine functions, two additional figures are provided, Figure-6A & 6B.   Each is a ten spoke, three-cycle Phi wheel, and spokes are separated by 36 degrees.   In these two figures, the trigonometric values of the angles (**n**) are displayed as vectors.

We have seen in the tables presented here how distillation and numerical expansion of the Fibonacci series leads to the creation of these unusual mathematical tables that exhibit the properties of numerical mirror imagery and symmetry.   The column and row summations, and the sums of certain gray shaded areas of symmetry, offer historically significant numbers that are repeatedly found throughout nature and in the structure of our universe.   Many of these numbers in the past were given religious significance and meaning in the old languages of Hebrew, Greek and Arabic, in systems that were concerned with the assignment of numbers to sounds and letters of their alphabets.   This was the science known as Gematria.

The Fibonacci series, and its associated Golden ratio Phi (φ), are an integral part of our natural world, and have been a controlling element in most all of humanity's great works of creation since the beginning of recorded time.

The cyclic rhythms of our universe are surely the basis of man's concept of *time*.   The orbital rotations of the heavenly

---

[1]   A theory or doctrine describing the natural order of the universe.

bodies which man observed required him to conceive a means to measure such movement. The "amount of movement" was most certainly based on the daily rising and setting of the sun and the seasonal changes that occurred before the long cycle was repeated. The adoption of **360** as the number of *"degrees"* for this annual movement was quite possibly a result of the number of days they counted in their early measurement of annual time. i.e.

The *"daily egress"* of the rising sun.

The choice of **360 degrees** per cycle has been accepted and used since the first early application in antiquity and remains in use today. Division of a circle into twenty equal angles, as shown by the twenty spoke, three cycle Phi (φ) wheel of Figure-6, illustrates why these angular numbers might have been chosen for use as linear measurements by the ancient artists and architects. It is apparently because so many of these same numbers or fractals thereof are found to be prevalent in nature, and notably most all of these numbers have trigonometric values that can be expressed in terms of Phi (φ). The exception being those angles called the cardinal points of the compass. i.e. ($90°$- $180°$- $270°$- $360°$).

Figure-6A and 6B display respectively three cycles of Sine and Cosine values. The numbers that end with a 4 or 6 have a trigonometric value of $\pm \varphi/2$, and numbers that end with a 2 or 8 have a trigonometric value of $\pm 1/(2\varphi)$.

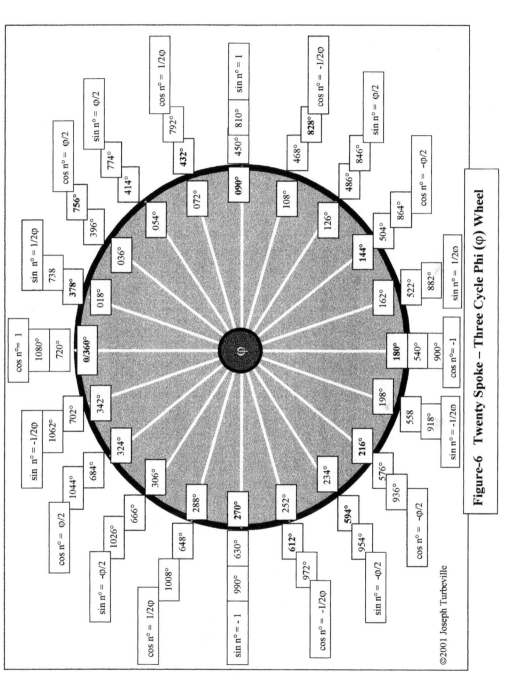

Figure-6 Twenty Spoke – Three Cycle Phi (φ) Wheel

©2001 Joseph Turbeville

51

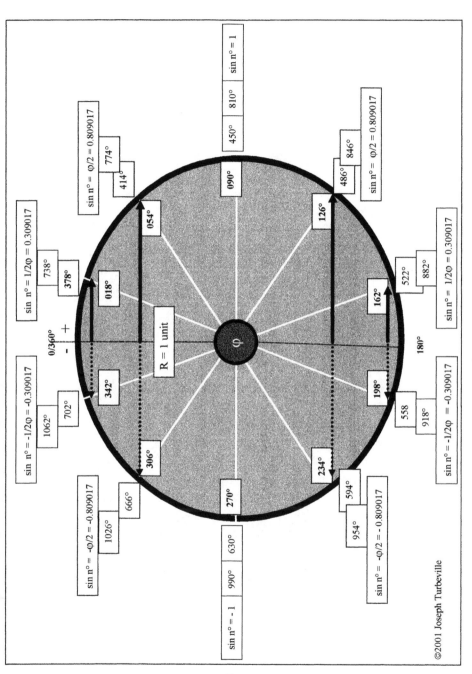

Figure-6A   Ten Spoke – Three Cycle – Sine – Phi (φ) Wheel

©2001 Joseph Turbeville

53

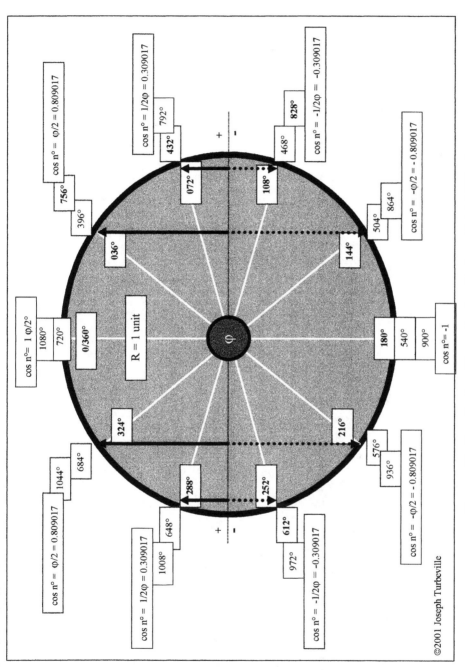

Figure-6B Ten Spoke - Three Cycle - Cosine - Phi (φ) Wheel

©2001 Joseph Turbeville

55

The following diagram (Fig-7) reveals a mathematical solution for the Golden Ratio (φ) and the numerical values for the vectors of Figure-6A & 6B.

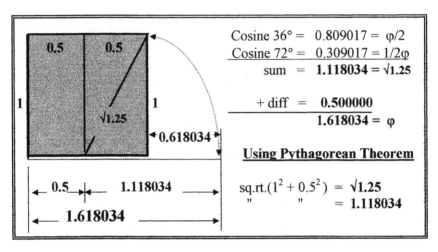

Figure-7   Golden Ratio (φ) Solution

A trigonometric Phi (φ) Function Rule is provided here to help identify those numbers (n) whose trigonometric values can be expressed as a function of Phi(φ).

1. If (n) is an integer divisible by 9, and (n) ÷ 360 contains one decimal place, (i.e., .1, .2, .3, .4, .6, .7, .8, .9), excluding (.0 & .5), then the Cosine (n) can be expressed as a function of Phi.
   i.e. cos (n) = $f(φ)$ If (n) ÷ 360 ends with (.0 or .5), then   cos (n) = ±1.

2. If (n) is an integer divisible by 9, and (n) ÷ 360 contains two decimal places, that are an odd multiple of (.05), (i.e., .05, .15, .35, .45, .55, .65, .85, .95) excluding (.25 & .75), then the Sine (n) can be expressed as a function of Phi.
   i.e.   sin (n) = $f(φ)$ If (n) ÷ 360 ends with (.25 or .75), then sin (n) = ±1

3. The numbers (n) that end with a 4 or 6 have a trig. function of ± φ/2
   The numbers (n) that end with a 2 or 8 have a trig. function of ± 1/(2φ)

Figure-8 Trigonometric Phi Function Rule

Trigonometric Phi (φ) Function Numbers in a Fibonacci Series Format

In order to make comparisons to the Phi (φ) function numbers shown on the Phi-wheels, to many of the same numbers shown in the Fibonacci series format of Table-4A, the reader should note the following:

□ Table-4A displays the Fibonacci series patterns in eight columns of gray scale for ease of observation and discussion.

□ Odd multiples of nine in Row-1 do not satisfy the Phi function identity rule and are marked (n/a), as not applicable.

□ Examination of Table-4A will reveal that none of the numbers outside of the first revolution on the 3-cycle Sine- wheel with the exception of **378** are present. The reader should note that Rows 3, 4, 5 & 6   only have numbers that satisfy the Cosine portion of the *Phi function identity rule.*

□ Columns 5, 10, 15, and 20 contain numbers that all end in zero, therefore according to the t*rigonometric Phi function identity rule* their trigonometric value is ± 1. The numbers that are on the second row of these four columns are termed the cardinal points of the compass, **90°, 180°, 270°, 360°.** These are named respectively East, South, West and North. Vectors normal to each of the four faces of the Great pyramid will have these exact compass bearings.

□ The numbers in Row-2 that correspond to the one through twenty sequence of Base numbers, are the same numbers that make up the initial cycle on the twenty-spoke Phi (φ) wheel in Part-2, Figure-6 (p51).

- Number **378** is the largest sine function number shown in Table-4A, and its trigonometric value is $0.3090017 = 1/(2\varphi)$. The half-baseline width of the Great pyramid is **378** feet as was shown in the section on *Surface Dimension Diagrams Utilizing Tabular Numbers*.

- The first three non-Fibonacci cells of Row-3 (the white cells), containing the numbers **144, 216 and 252** have a sum of **612,** which, expressed in feet, is the apothem of the Great pyramid.

- The next group of non-Fibonacci cells in row three (the four adjacent white cells), i.e. **324, 360, 396** and **432**, have an average value **378**, half the base width of the Great pyramid when measured in feet.

- The last group of non-Fibonacci cells in row three (the seven adjacent white cells), i.e. **504, 540, 576, 612, 648, 684,** and **720** have an average value of **612**, expressed in feet, is the apothem of the Great pyramid.

- A few of the major tabular numbers that have been under discussion in the development of this thesis, are the base-line measurements listed in column number twenty-one of Table-4A. (it is actually the 22nd column)

Figure-8A below provides a connection to the pyramid Pi(π) value.

Doubling $\Downarrow$		π	Column 21 of Table 4A $\Downarrow$		
$1900800 \div 100 = 19008 \Rightarrow$	**19008** $\div 22/7 =$	$6048 \Rightarrow$	**6048** ft =	Khufu mile (p. 25)
$1900800 \div 200 = 9504 \Rightarrow$	**9504** $\div 22/7 =$	$3024 \Rightarrow$	**3024** ft =	Khufu's perimeter
$1900800 \div 400 = 4752 \Rightarrow$	**4752** $\div 22/7 =$	$1512 \Rightarrow$	**1512** ft =	Khufu's (½) " .
$1900800 \div 800 = 2376 \Rightarrow$	**2376** $\div 22/7 =$	$756 \Rightarrow$	**756** ft =	Khufu's base width
$1900800 \div 1600 = 1188 \Rightarrow$	**1188** $\div 22/7 =$	$378 \Rightarrow$	**378** ft =	Khufu's (½) width
$1900800 \div 3200 = 594 \Rightarrow$	**594** $\div 22/7 =$	$189 \Rightarrow$	**189** ft =	Khufu's (¼) width

Figure-8A   Pyramid Pi (π_p) Connection

The bottom row of numbers in this table form a limited set of baseline numbers and the gray cells represent the first eight terms of the Fibonacci Series. Column base multipliers (9, 18, 36, 72, 144 & 288) form row 23 ascending columns of doubled numbers, many of which are subjects of this work.

6	288	576	864	1152	1440	1728	2016	2304	2592	2880	3168	3456	3744	4032	4320	4608	4896	5184	5472	5760	6048	6336
5	144	288	432	576	720	864	1008	1152	1296	1440	1584	1728	1872	2016	2160	2304	2448	2592	2736	2880	3024	3168
4	72	144	216	288	360	432	504	576	648	720	792	864	936	1008	1080	1152	1224	1296	1368	1440	1512	1584
3	36	72	108	144	180	216	252	288	324	360	396	432	468	504	540	576	612	648	684	720	756	792
2	18	36	54	72	90	108	126	144	162	180	198	216	234	252	270	288	306	324	342	360	378	396
1	9	18	27	36	45	54	63	72	81	90	99	108	117	126	135	144	153	162	171	180	189	198
B	1	2	3	4	5	6	7	8	9	10	11	12	13	14	15	16	17	18	19	20	21	22

Notes Applicable to Table-1.

- The first three white cell numbers in row # 3 have a sum of 612, which is the **apothem** length of the Great pyramid measured in feet. i.e. (144 + 216 + 252) = 612 feet.

- The next four white cell numbers in row # 3, have a sum of 1512 which is half the baseline circumference of the Great pyramid measured in feet. (324 + 360 + 396 + 432) = 1512 feet.

- The next three hi-lighted numbers in row # 3, 612, 720 and 756 are respectively the apothem length, the corner-line length and the base width of the Great Pyramid. i.e. **apothem = 612** feet, **corner-line = 720** feet, **baseline width = 756** feet.

- True compass bearings for the four faces of the Great Pyramid are highlighted in row # 2. East, South, West and North. --- 90° - 180° - 270° - 360° respectively.

TABLE-4   A Sequential Assembly of Multiples of Nine Displayed in an Eight Column Fibonacci Series Format.

Row																					
Sign RH6	cos 288	cos 576	cos 864	cos 1152	cos 1440	cos 1728	cos 2016	cos 2304	cos 2592	cos 2880	cos 3168	cos 3456	cos 3744	cos 4032	cos 4320	cos 4608	cos 4896	cos 5184	cos 5472	cos 5760	cos 6048
f(φ)	1/2φ	-φ/2	-φ/2	1/2φ	+1	1/2φ	-φ/2	-φ/2	1/2φ	+1	1/2φ	-φ/2	-φ/2	1/2φ	+1	-φ/2	-φ/2	-φ/2	1/2φ	+1	1/2φ
Rev	0.8	1.6	2.4	3.2	4.0	4.8	5.6	6.4	7.2	8.0	8.8	9.6	10.4	11.2	12.0	12.8	13.6	14.4	15.2	16.0	16.8
Sign RH5	cos 144	cos 288	cos 432	cos 576	cos 720	cos 864	cos 1008	cos 1152	cos 1296	cos 1440	cos 1584	cos 1728	cos 1872	cos 2016	cos 2160	cos 2304	cos 2448	cos 2592	cos 2736	cos 2880	cos 3024
f(φ)	-φ/2	1/2φ	1/2φ	-φ/2	+1	-φ/2	1/2φ	1/2φ	-φ/2	+1	-φ/2	1/2φ	1/2φ	-φ/2	+1	-φ/2	1/2φ	1/2φ	-φ/2	+1	-φ/2
Rev	0.4	0.8	1.2	1.6	2.0	2.4	2.8	3.2	3.6	4.0	4.4	4.8	5.2	5.6	6.0	6.4	6.8	7.2	7.6	8.0	8.4
Sign RH4	cos 72	cos 144	cos 216	cos 288	cos 360	cos 432	cos 504	cos 576	cos 648	cos 720	cos 792	cos 864	cos 936	cos 1008	cos 1080	cos 1152	cos 1224	cos 1296	cos 1368	cos 1440	cos 1512
f(φ)	1/2φ	-φ/2	-φ/2	1/2φ	+1	1/2φ	-φ/2	-φ/2	1/2φ	+1	1/2φ	-φ/2	-φ/2	1/2φ	+1	1/2φ	-φ/2	-φ/2	1/2φ	+1	1/2φ
Rev	0.2	0.4	0.6	0.8	1.0	1.2	1.4	1.6	1.8	2.0	2.2	2.4	2.6	2.8	3.0	3.2	3.4	3.6	3.8	4.0	4.2
Sign RH3	cos 36	cos 72	cos 108	cos 144	cos 180	cos 216	cos 252	cos 288	cos 324	cos 360	cos 396	cos 432	cos 468	cos 504	sine 540	cos 576	cos 612	cos 648	cos 684	cos 720	cos 756
f(φ)	φ/2	φ/2	-1/2φ	-φ/2	-1	-φ/2	-1/2φ	1/2φ	φ/2	+1	φ/2	1/2φ	-1/2φ	-φ/2	-1	-φ/2	-1/2φ	1/2φ	φ/2	1	1/2φ
Rev	0.1	0.2	0.3	0.4	0.5	0.6	0.7	0.8	0.9	1.0	1.1	1.2	1.3	1.4	1.5	1.6	1.7	1.8	1.9	2.0	2.1
Sign RH2	sine 18	sine 36	sine 54	cos 72	sine 90	cos 108	sine 126	cos 144	sine 162	cos 180	sine 198	cos 216	sine 234	sine 252	sine 270	cos 288	sine 306	cos 324	sine 342	cos 360	sine 378
f(φ)	φ/2	φ/2	φ/2	1/2φ	+1	-1/2φ	φ/2	-φ/2	1/2φ	-1	-1/2φ	-φ/2	1/2φ	-1/2φ	-1	1/2φ	-φ/2	φ/2	-1/2φ	+1	1/2φ
Rev	0.05	0.1	0.15	0.2	0.25	0.3	0.35	0.4	0.45	0.5	0.55	0.6	0.65	0.7	0.75	0.8	0.85	0.9	0.95	1.0	1.05
Sign RH1	n/a 9	sine 18	n/a 27	cos 36	n/a 45	sine 54	n/a 63	cos 72	n/a 81	sine 90	n/a 99	cos 108	n/a 117	sine 126	n/a 135	cos 144	n/a 153	sine 162	n/a 171	cos 180	n/a 189
f(φ)	n/a	1/2φ	n/a	1/2φ	n/a	φ/2	n/a	1/2φ	n/a	+1	n/a	-1/2φ	n/a	φ/2	n/a	-φ/2	n/a	1/2φ	n/a	-1	n/a
Rev	0.025	0.05	0.075	0.1	0.125	0.15	0.175	0.2	0.225	0.25	0.275	0.3	0.325	0.35	0.375	0.4	0.425	0.45	0.475	0.5	0.525
Base	1	2	3	4	5	6	7	8	9	10	11	12	13	14	15	16	17	18	19	20	21

Trigonometric Phi function rules

1. If (n) is an integer divisible by 9, and (n) ÷360 contains one decimal place, (i.e., .1, .2, .3, .4, .6, .7, .8, .9), excluding (.0 or .5), then the Cosine of (n) can be expressed as a function of Phi. -- i.e. cos.(n) = f (φ). ---- If (n) ÷360 ends with (.0 or .5), then cos.(n) = ± 1.

2. If (n) ÷360 has two decimal places and is an odd multiple of (.05), (i.e., .05, .15, .35, .45, .55, .65, .85, .95), excluding (.25 & .75), then the Sine of (n) can be expressed as a function of Phi. -- i.e. sin.(n) = f (φ). ---- If (n) ÷360 ends with (.25 or .75), then sin.(n) = ±1.

3. The numbers (n) that end with a 4 or 6 have a trig function of ± φ/2. & The numbers (n) that end with a 2 or 8 have a trig function of ± 1/(2φ).

TABLE-4A   The Sequential Multiples of Nine and Application of the Trigonometric Phi Function (φ) Rules

Velocity Attained Falling in a Hypothetical Hole Through the Center of the Earth. - 25920 ft/sec.

A classic problem that has often been presented to students of Physics, and one that is totally impossible and entirely hypothetical, concerns the dropping of a ball into a "hole" that has been bored straight down through the center of the Earth. The gravitational factor drops linearly from 32.1327 ft/sec^2 at the Earth's surface (~ 30 deg. Latitude) down to zero at the Earth's center. Thirty degrees was chosen because it is the approximate latitude of the Great Pyramid, a primary subject of this thesis. The average value for g therefore is $(g_{avg}) = 16.0664$ ft/sec^2.

Equating the *potential energy* that the ball would have at the moment it is dropped, to the *kinetic energy* it would have as it reaches the center of the Earth, we can determine its maximum velocity. Thereafter, it would begin to lose speed and come to a stop just as it reaches the surface on the other side of the Earth. Unless the ball is removed from the "hole", it would fall back in the other direction and repeat the process again and again, as an oscillation, similar to a mass on a spring. This model assumes no energy loss due to air friction.

(Potential Energy) **PE** = (Kinetic Energy) **KE**
mass x g_{avg} x height = ½ mass x velocity^2
velocity^2 = 2 x g_{avg} x Radius $_{Earth}$
velocity = $\sqrt{2}$ x 16.0664 ft/sec^2 x 20,908,800 ft.)
velocity = **25920** ft./sec

It is extremely interesting that the numerical value for the maximum velocity in this hypothetical problem is identical to the tabular value for the period for the Earth's ecliptic cycle of **25920** years.

The Critical Speed for a Hypothetical Low Level Earth Orbit

In our universe, the force of attraction between any two bodies is directly proportional to the product of their masses and inversely proportional to the square of the distance between body centers.

At the Earth's surface the gravitational force a body experiences depends on its geographical location because the value of the Earth's radius (R_{Earth}) increases slightly as one moves toward the equator. In physical terms, the gravitational force exerted on a body of mass (m) at or near the surface is:

$$Force = weight = mg = GmM_{Earth}/(R_{Earth})^2$$

$$\cancel{m}g = Gm M_{Earth}/(R_{Earth})^2 \qquad G = Universal\ Constant$$

$$g = GM_{Earth}/(R_{Earth})^2 = gravitational\ acceleration$$

The distance that an object of mass (m) travels when launched horizontally from some high point close to the Earth's surface depends on the speed it has at launch and the air resistance it encounters The gravitational force acts at a right angle to the horizontal path of the object, (i.e. toward the Earth's center) and is the force that pulls this object toward the Earth. The faster it is traveling, the greater the horizontal distance it will traverse before being pulled to Earth. There is a critical speed at which an object can maintain its initial height without falling to Earth. This problem becomes hypothetical when one realizes that such a speed, close to the earth's surface is impossible because of air friction. It causes excessive temperatures that could burn up a projectile. In the absence of any atmosphere, it would be able to orbit in a great circle path just above the Earth's surface. The critical speed for such a hypothetical low-level great circle orbit is given by the equation for centripetal acceleration, $(a)_c = v^2/R$. When $(a)_c$ is equal to (g), a great circle orbit could be maintained.

$$Velocity^2 = g \times R_{Earth} = (32.1327\ ft/sec^2)^* \times (3960 s.mi \times 5280\ ft/s.mi)$$

$$Hypothetical\ Velocity = \sqrt{(671{,}856{,}197.8\ ft^2/sec^2)} = \mathbf{25920}\ ft/sec.$$

"g" at $31°$ latitude = 32.1327 ft/sec^2 -Handbook of Chemistry and Physics - 62nd Ed. 1981–82

The "hypothetical critical speed" required for a low-level great circle Earth orbit, (**25920** ft/sec), and the maximum velocity (**25920** ft/sec), attained by a ball dropped in a "hypothetical hole" drilled through the center of the Earth, both previously discussed, share something other than identical speeds. The hypothetical Earth orbital time and the oscillatory period of the ball in the "hypothetical hole" form a square root ratio approximating the Golden Ratio Phi (φ). i.e.

- The hypothetical great circle orbital time (t) for a low-level Earth orbit is equal to the average circumference of the Earth (in feet) divided by the required speed (**25920** ft/sec).
  Orbit Time = π_p^*(7920 s.mi)(5280 ft/s.mi)/(25920 ft/sec) = 5070.5 sec.

- The oscillatory period (T) for the falling ball is equal to twice the Earth's mean diameter divided by the ball's mean speed.
  Period (T) = 2(7920 s.mi.)(5280 ft/s.mi.)/(25920/2) ft/sec. = 6453.3 sec.

- The ratio of the oscillatory period (T) to the orbital time (t) is approximately the square root of Phi (φ) i.e. The Golden Mean.

- Oscillatory period / Orbital time = 6453.3sec./ 5070.5 sec.= 1.2727
  ~ The square root of the Golden Mean = 1.27202

Ratio Squared = $(1.2727)^2$ = 1.6198 ≅ Phi (φ) (0.11 % error)
True value of Golden Mean = 1.618033989 = Phi (φ)

---

* The value used for Pi is what has been termed the Pyramid Pi (22/7) by the author.

It is important to point out that this hypothetical example is also valid for the Earth's moon or any other planetary body, as long as the associated gravitational constant for the body is used. In any case, the critical speed for a low-level orbit and the maximum speed for the hypothetical oscillating "ball in the hole" will be identical. In the Earth example, the speed was determined to be **25920** ft/sec., *a numerical value of historical significance.*

The Earth's moon is used here as an additional example. The gravitational constant (g') of the moon, at or near its surface, is directly proportional to the moon's mass and inversely proportional to the square of its radius. Newton's gravitation term (G), is the proportionality constant.

$mg' = GmM_{moon} / R^2 = (6.673 \times 10^{-11} \, Nm^2 \, kg^{-2})(7.36 \times 10^{22} \, kg)/ (1.73809152)^2$
Moon's $g' = 1.625749607 \, m/s^2 = 5.3338 \, ft/s^2 \cong (1/6)$ Earth's (g) value.

The critical speed for low-level moon orbit:
As discussed previously, the critical speed to remain in a low-level great circle orbit requires that the gravitational constant (g') be equal to the centripetal acceleration (a_c).

$g' = a_c = V^2/R \Rightarrow V^2 = g'R \Rightarrow V = \sqrt{(g'R)} = \sqrt{(5.3338 \, ft/s^2 \times 5,702,400 \, ft)}$
$V = 5515 \, ft/s$
Orbital Period (t) = Circum./ V = $(\pi_P \times 2160 \, s.mi \times 5280 \, ft/s.mi.) / 5515 \, ft/s.$
Orbital Period (t) = 6499.3 sec. $\cong$ 6500 sec.

The maximum speed $(V)_{max}$ attained by the ball falling through a hypothetical "hole through in moon" occurs just as it passes through the center of the Moon.

Here again, we assume no frictional loss of energy and equate the potential energy (PE) at the top of the hole, to the kinetic energy (KE), it would have at the Moon's center, and determine (V_{max}).

$PE = KE \Rightarrow m \cdot R(g'_{avg}) = \frac{1}{2} \cdot m \cdot V^2_{max} \Rightarrow V^2_{max} = 2(g'_{avg})R$

$V_{max} = \sqrt{(2(g'_{avg})R)}$
$V_{max} = \sqrt{(2 \times 5.3338/2 \text{ ft/s}^2 \times 5,702,400 \text{ ft})} = 5515 \text{ ft/s}$
$V_{avg} = V_{max}/2 = 2,757.5 \text{ ft/sec.}$

Oscill. Period (T) = 2 Dia./ V_{avg} = 2(2160 mi x 5280 ft/ mi)/2,757.5 ft/sec.
Oscill. Period (T) = 8271.8 sec.

RATIO: Oscill. Period /Orbital Period = 8271.8 s/6499.3 s = 1.27272 = $\sqrt{\varphi}$

This value 1.272727 · · · · when squared, was earlier[*] shown to be approximately equal to the Golden ratio (φ), with an error of 0.11%. It was termed *Pyramid Phi* (φ_1) by this author because of its strong connection to *Pyramid Pi* (π_p) and the tabular sums from Table-1 (756 and 594) that form the ratio below.

$\sqrt{\varphi_1} \equiv 4/(22/7) = 756/594 = 1.272727$

It is also closely approximated by the square root of two other tabular sums from Table-1. i.e. $\sqrt{(612/378)} = 1.272418$

---

* See p.22

The Nineveh Number, The Alpha Numeric Value of Nine, and The Numeral 9

My introduction to the number of Nineveh (1.959552×10^{14} sec.) from my readings of Maurice Chatelain[7] and Charles W. Johnson[5] evoked in me a heightened interest in our ancient past and further 'nudged' me back in time to try and make new connections with some of the historical tabular numbers presented in my earlier writings.

One of my first ideas was to utilize the "*ratio of nines*" which Gary Val Tenuta[2] the author of *The Secret of Nine*, had just recently discussed with me. The "*ratio of nines*" is actually a ratio of the alphanumeric value (ANV) of the English word NINE, divided by the numeral 9. i.e.

$$\text{ANV of } (N+I+N+E)/9 = (14+9+14+5)/9 = 42/9 = 4.6666 \qquad \text{-- eq.-1}$$

Needless to say, the first four letters in the word *Nineveh* prompted me to take a look at the alphanumeric approach in my examination of the "Great Constant".

$$1.959552 \times 10^{14} \text{ sec } \times (42/9) = 9{,}144{,}576 \times 10^{14} \text{ sec} = (3024 \times 10^4 \sqrt{\text{sec}})^2 \qquad \text{-- eq.-2}$$

Then, looking at the Nineveh number times the reciprocal of the above ratio we obtain:
$$1.959552 \times 10^{14} \text{ sec } \times (9/42) = 4.199040 \times 10^{13} \text{ sec} = (648 \times 10^4 \sqrt{\text{sec}})^2 \qquad \text{--- eq.-3}$$
The product of the two roots is equal to the Nineveh number.

$$648 \times 10^4 \sqrt{\text{sec}} \times 3024 \times 10^4 \sqrt{\text{sec}} = 1.959552 \times 10^{14} \text{ sec} = \text{Nineveh Number- eq.- 4}$$

---

[2] Gary Val Tenuta - *The Secret of Nine* - self published - CodeUFO@aol.com

Recalling the tabular value of **3024** feet as the probable design baseline perimeter of the Great Pyramid, and being aware of the generally held belief that the pyramid was designed as a model of the Earth's Northern hemisphere, suggests that a factor might exist that would allow conversion of this equation from units of time to units of length, and provide a measure of the Earth's equatorial circumference, as well as a scale factor for the size of the pyramid as compared to the Earth.

$$\frac{648 \times 10^4 \sqrt{sec} \times 3024 \times 10^4 \sqrt{sec}}{1.5 \times 10^2 \sqrt{sec} \times 10^4 \sqrt{sec/ft}} = \frac{1.959552 \times 10^{14} \ sec}{1.5 \times 10^6 \ sec/ft} \qquad \text{--- eq.-5}$$

$$\underline{43200} \times \underline{3024 \ ft} = 130,636,800 \ ft = 24,742 \ s.mile \quad \text{(0.6\% error)} \qquad \text{--- eq.-6}$$

Earth / GP scale x GP Base perimeter = Earth's circumference.   -- eq.-7

Note: 24,742 s. mile ≡ 21600 Khufu mile

In addition, by applying the ancient practice of doubling, as suggested by Johnson[5], the fractal 648 when doubled twice becomes 2592, and the Nineveh fractal 1959552 when divided by the 2592 equals 756. Recall, the baseline width of Khufu's pyramid is 756 feet, and the period of the ecliptic cycle is 25920 years.

The conversion factor (1.5×10^6 sec/ft) represents the time required for a point on the Earth's equator to make one full axial rotation (130,636,800 ft.) in the same time frame that it takes to make 240 ecliptic rotations. This ecliptic perturbation occurs because both the sun and the moon exert a pull on the Earth's equatorial bulge, causing its rotational axis to wobble and slowly precess around the axis of the ecliptic at a rate of one degree every 72 years or 25920 years per ecliptic cycle. The critical reader might observe that the Nineveh number when divided by 240 produces an ecliptic period of 25890.41096 year (365 d/y) that differs from my tabular data (25920 yr.) by just 30 years. (An error of 0.1 %)

An Angular Perspective of Leonardo da Vinci's Vitruvian Man: Utilizing the Phi (φ) Function Identity Rule of Natural Symmetry

The famous drawing of the *Vitruvian Man* visually defines the nature of man's physical proportions in terms of the geometric bounds of the square and the circle. This dual projection of man into the overlaid square and circle may additionally provide the viewer with a sense of motion. The projection into the square with his arms out-stretched horizontally, touching the sides of the square, and his legs together relays to the viewer the fact that the man's arm-span is equal to his height.

The man's projection into the circle, which is tangent to the square at the baseline, provides the reader with a pair of historically significant, *numerical mirror image* angles adjacent to the vertical line drawn through the body center. 72° and 27°.

Aware of the abundance of writings over the years that dwell on the *Golden ratio Phi(φ)*, (*1.618034···*) and the bodily proportions available in this drawing, the author has elected to mention only the *vertical distance from the baseline to the naval, divided by the distance from the naval to the top of the head, a ratio approximately equal to Phi(φ)*. The remainder of this section is a discussion of the angles formed by the arms and legs and their trigonometric values that may be expressed in terms of Phi, as allowed by the *Trigonometric Phi function identity rule.*[*]

The first major historically significant number discussed is the angle extended by the outstretched arms over the head of the man in the circle. Contact is made with the fingertips at the edge of the circle and the angle formed is **144°**. Applying the *Phi function identity rule* to this number we divide **144°** by **360°** and obtain **0.4** parts of a revolution. The single decimal place signifies that the Cosine of **144°** can be expressed in terms of Phi (φ). i.e.

Cosine **144°** = -0.8090 = -φ/2

---

[*] See Appendix of this section.

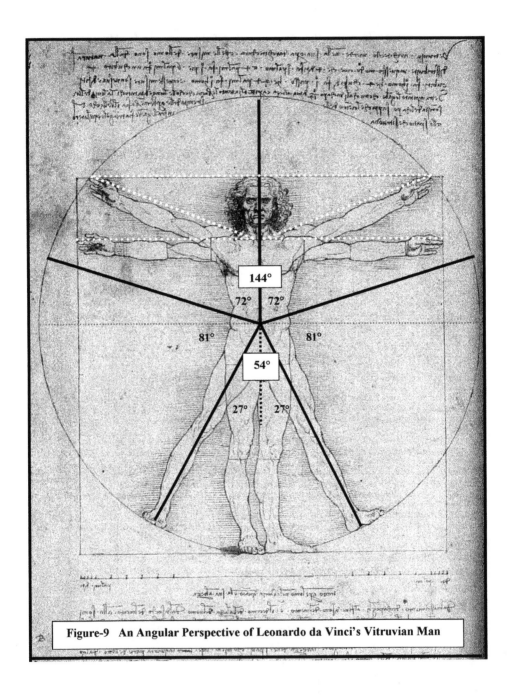

Figure-9 An Angular Perspective of Leonardo da Vinci's Vitruvian Man

The second historically significant number is the angle **54°** that is extended by the open leg stance of the Vitruvian man in the circle. Applying the *trigonometric Phi function identity rule* to this number, **0.15** parts of a revolution is obtained. The two decimal places signify that the Sine **54°** can be expressed in terms of Phi (φ).

    i.e. Sine **54°** = 0.8090 = $\varphi/2$.

Note at this point that the half-angles of **144°** and **54°** are respectively **72°** and **27°**, a unique *numerical mirror image* pair that we will further examine. As discussed earlier, these historically significant numbers are found from time to time to take on other units of the Imperial system of measure. e.g.

- The angle **72°** and its adjacent angle **81°** have a sum of **153°**. Historically, number **153** relates to the quantity of fishes caught in the net of the disciple Simon Peter in the gospel of St. John. The Cosine of **72°** = 0.3090$\cdots$ = $1/(2\varphi)$.

  Note, the 153rd course of the Great Pyramid is at an average height of 4379.85 inch [2] = 365 feet. It is often cited in reference to the number of days in a year. i.e. 365 d/y.

- The two **27°** angles and their adjacent **81°** angles have a total sum of **216°**. The ratio presented by the arms of the Vitruvian man with the division of the circle into two parts is **144 : 216**.

The ratio **144:216 = 2:3** is important here in that Socrates in a discussion of musical harmony in Plato's *The Marriage Allegory (Republic)*[8] comments that the *"human male"*, prime number five,

[2] *W.M.F.Petrie - The Pyramids and Temples of Gizeh – Course Data – Published London-1883*
[8] *Reference - E.G.McClain –"The Pythagorean Plato: Prelude to the Song Itself"- p.23 ISBN 0-89254-010-9 –1984 - Publisher Nicolas-Hays, Inc.- York Beach, Maine-03910*

enters harmonic theory as an *arithmetic mean* within the perfect fifth of 2:3 – expanded to 4:5:6 to avoid fractions.

Did Leonardo da Vinci select the angular ratio 2:3 for placement of his Vitruvian "human male" in the circle, or was it just an unavoidable fact that was by nature the only possible choice? It is the author's belief that the angular perspective presented here, concerning the use of angles whose trigonometric functions can be expressed in terms of Phi(φ), offers further evidence of nature's influence on the great works of man.

Many angular images found in nature satisfy the Phi (φ) function identity rule, and such angles are found in many of man's creative works. These angles may occur, unknown to the artist, because of physical restrictions that nature places on the artist's subject. Just such a limitation is displayed by the angles required for the Vitruvian man to make four-point contact with his circle, when his navel is considered the focus of that circle. The viewer may also be unaware that this limitation is possibly a result of a biological Phi function requirement, unless they are told of it.

Listed below are the major angles and various angular combinations from the drawing of the Vitruvian man.

- $144° = 72° + 72°$ Cosine $144° = -0.8090 = -\varphi/2$
- $72°$ Cosine $72° = 0.3090 = 1/(2\varphi)$
- $54° = 27° + 27°$ Sine $54° = 0.8090 = \varphi/2$
- $27°$ $27°$ not a Phi φ function number
- $81°$ $81°$ not a Phi φ function number

- 144° + 54° = 198°   Sine 198° = -0.3090 = -1/(2φ)
- 27° + 81° = 108°   Cosine 108° = -0.3090 = -1/(2φ)
- 81° + 81° = 162°   Sine 162° = 0.3090 = 1/(2φ)
- 72° + 81° = 153°   153° not a Phi (φ) function number
- 72° + 27° = 99°   99° not a Phi (φ) function number

Note: All odd numbers and sums that are odd numbers fail the Phi function rule. However, when an odd numbered angle is viewed as a *mirror image pair*, the double-angle is seen as an even number and may possibly obey the Phi function selection rule. The two angles, 81°opposite 81°, form a *mirror image pair* that has a sum of 162°. This satisfies the Phi function rule just as does the angle 27° opposite 27° that form the 54°open leg stance of the Vitruvian man.

---

1. If (n) is an integer divisible by 9, and (n) ÷ 360 contains one decimal place, (i.e., .1, .2, .3, .4, .6, .7, .8, .9), excluding (.0 & .5), then the Cosine (n) can be expressed as a function of Phi. ---- i.e. cosine (n) = $f(\varphi)$. If (n) ÷ 360 ends with (.0 or .5), then Cosine (n) = ± 1.

2. If (n) is an integer divisible by 9, and (n) ÷ 360 contains two decimal places that are an odd multiple of (.05), (i.e., .05, .15, .35, .45, .55, .65, .85, .95) excluding (.25 & .75), then the Sine (n) can be expressed as a function of Phi. i.e. sine (n) = $f(\varphi)$. If (n) ÷ 360 ends with (.25 or .75), then Sine (n) = ± 1.

3. The numbers (n) that end with a 4 or 6 have a trig. function of ± φ/2. The numbers (n) that end with a 2 or 8 have a trig. function of ± 1/(2φ).

Figure-8   Trigonometric Phi Function Rule

The rhythm of this natural series of numbers *(Fibonacci series)* is found in the leaf arrangement on plant stems, various pedal counts, and the seed head arrangement of sunflowers and pinecones. This is a subject found in the study of botany called *phyllotaxis*. Botanical literature concerning the subject is widely available both online and in public libraries.

The plant selected for discussion is a common spring weed found in the United States called Daisy Fleabane and has a five-leaf stem rotation pattern. The two specimens selected for this discussion both have rotation angles that closely approximate Phi function angles. Additionally they were selected because one of them is identical to a plant discussed on the Fibonacci website of Prof. Ron Knott*. Figures-10, 10A & 11 reveal two different five-leaf rotation patterns that nature has bestowed on them in order that all leaves equally shares all available sunlight and water.

The reader should carefully note the number of rotations required for a repetitive five-leaf rotation pattern, both clockwise and counter clockwise, then note that the rotation angle of each leaf satisfies the trigonometric Phi function selection rule, thus the numerical value can be expressed in terms of Phi.

---

*   http://www.mcs.surrey.ac.uk/Personal/R.Knott/Fibonacci/fibnat.html#plants

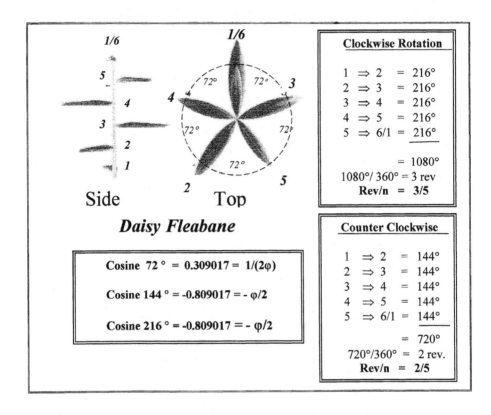

Figure-10 Computer Generated 5-Leaf Pattern with Equal Angles,
Phi (φ) Function Rotations and Ascending Leaf Positions.

Figures-10 and 10A are displays of a similar plant, with the exception that it was growing on a steep embankment that caused the leaf spacing to open much wider at the point where it faced the embankment. The angles on this computer-generated figure are approximated because of the difficulty in getting exact measurements due to the spindly nature of the leaves.

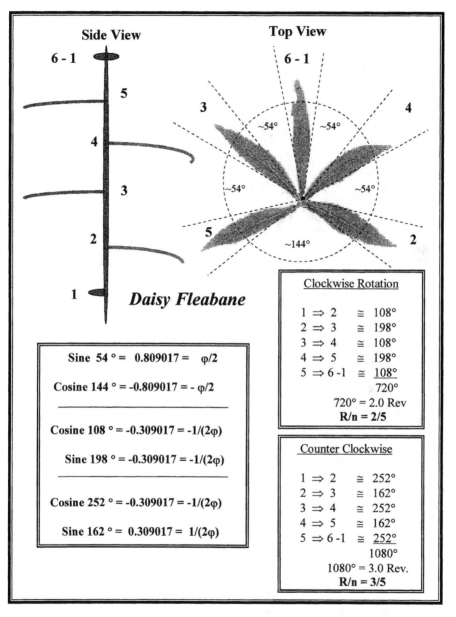

Side View

6 - 1

5

4

3

2

1

Daisy Fleabane

Top View

6 - 1

3

4

~54° ~54°

~54° ~54°

5

~144°

2

Sine 54 ° = 0.809017 = φ/2

Cosine 144 ° = -0.809017 = - φ/2

Cosine 108 ° = -0.309017 = -1/(2φ)

Sine 198 ° = -0.309017 = -1/(2φ)

Cosine 252 ° = -0.309017 = -1/(2φ)

Sine 162 ° = 0.309017 = 1/(2φ)

Clockwise Rotation

1 ⇒ 2   ≅ 108°
2 ⇒ 3   ≅ 198°
3 ⇒ 4   ≅ 108°
4 ⇒ 5   ≅ 198°
5 ⇒ 6 -1 ≅ 108°
            720°
720° = 2.0 Rev
R/n = 2/5

Counter Clockwise

1 ⇒ 2   ≅ 252°
2 ⇒ 3   ≅ 162°
3 ⇒ 4   ≅ 252°
4 ⇒ 5   ≅ 162°
5 ⇒ 6 -1 ≅ 252°
            1080°
1080° = 3.0 Rev.
R/n = 3/5

**Figure-10A   Computer Generated 5-Leaf Alternating Phi Function
Angles of Rotation and Ascending Leaf Positions**

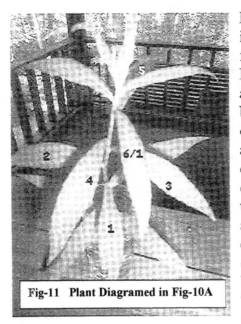

Fig-11 Plant Diagramed in Fig-10A

In summary, many angular images and growth patterns found in nature satisfy the Phi function identity rule, and as such, they may also be found in many of man's creative works of art. Such angles may occur in a work of art, unknown to the artist, due to physical restrictions that nature has placed on the artist's subject. One such example is found in the drawing by Leonardo da Vinci of The *Vitruvian Man*. The Phi (φ) function angles produced by the four-point contact of the man's arms and legs within the circle, is a result of a biological trigonometric Phi function requirement, that may have been unknown to the artist.

1. If (n) is an integer divisible by 9, and (n) ÷ 360 contains one decimal place, (i.e., .1, .2, .3, .4, .6, .7, .8, .9), excluding (.0 & .5), then the Cosine (n) can be expressed as a function of Phi. ---- i.e. cosine (n) = $f(\varphi)$.
   If (n) ÷ 360 ends with (.0 or .5), then Cosine (n) = ±1.

2. If (n) ÷ 360 has two decimal places which is an odd multiple of (.05), (i.e., .05, .15, .35, .45, .55, .65, .85, .95) excluding (.25 & .75), then the Sine (n) can be expressed as a function of Phi. --- i.e. sine (n) = $f(\varphi)$.
   If (n) ÷ 360 ends with (.25 or .75), then Sine (n) = ±1.
   Note:
3. The numbers (n) that end with a 4 or 6 have a trig. function of ± φ/2.
   The numbers (n) that end with a 2 or 8 have a trig. function of ± 1/(2φ).

Figure-8   Trigonometric Ph (φ) Function Identity Rule

Major Design Elements of the Great Pyramid of Giza Discovered in Gematrian Researcher's Alpha Number Table

The Alpha Number letter sets of Gary Val Tenuta[2] and their mirror image numbers allow the formation of matrices that amazingly contain the major external design elements of the Great Pyramid.

Ratios are formed from some of the tabular numbers that provide very close approximations of Pi (π) and the Golden ratio Phi (φ). The trigonometric value of some of the matrix numbers may also be expressed in terms of Phi (φ).

Tenuta's discovery concerning the *three, four and five letter* alphanumeric sets provided an important key in his work, and enabled this author to find new positive links to the *"numbers of nature"*.

The Alpha Number Table ©2001 Gary Val Tenuta			
Alpha Number	Alphanumeric Value "ANV"	Reduced Value	Set Identification
ZERO	64	1	(Stand Alone)
ONE	34	7	3-Letter Set
TWO	58	4	3-Letter Set
THREE	56	2	5-Letter Set
FOUR	60	6	4-Letter Set
FIVE	42	6	4-Letter Set
SIX	52	7	3-Letter Set
SEVEN	65	2	5-Letter Set
EIGHT	49	4	5-Letter Set
NINE	42	6	4-Letter Set

[2] Gary Val Tenuta – CodeUFO@aolcom

COLORS (Grouped and Summed for Analysis)

ZERO	64	1	(Stands Alone)
ONE	34	7	3-Letter Set
TWO	58	4	3-Letter Set
SIX	52	7	3-Letter Set
126	144	18	
FOUR	60	6	4-Letter Set
FIVE	42	6	4-Letter Set
NINE	42	6	4-Letter Set
459	144	18	
THREE	56	2	5-Letter Set
SEVEN	65	2	5-Letter Set
EIGHT	49	4	5-Letter Set
378	170	8	

Analysis of the Ascending and Summed Alpha-Numeric Color Groups
and
Their Historical Significance as Major Elements of the Great Pyramid

3-Letter Set	4-Letter Set	5-Letter Set
ONE – TWO – SIX	FOUR – FIVE – NINE	THREE – SEVEN – EIGHT
126	459	378

- Matrices of these number sets and their mirror numbers follow. Some numbers have an (s) or (c) displayed in front of them, which indicates that their trigonometric sine or cosine values may be expressed in terms of the Golden ratio Phi(φ) = 1.618034.

$$\text{e.g.} \quad \text{(s)} \quad \text{Sine } \mathbf{126} = +0.809016994 = (\varphi)/2$$
$$\text{(c)} \quad \text{Cosine } \mathbf{612} = -0.309016994 = 1/(2\ \varphi)$$

Matrix #	Mirror #	diff	Matrix #	Mirror #	diff	Matrix #	Mirror #	diff
s 126	621 →	495	459	s 954 →	495	s 378	873→	495
c 612	c 216 →	c 396	945	549 →	c396	837	s 738→	99
261	s 162 →	99	s 594	495 →	99	783	387→	c396
999	999	990	s 1998	s 1998→	990	s 1998	s 1998	990

- The trigonometric numbers of each of the three sets, when added, produce the following historically significant numbers.

Trig. #	Mirror #	diff	Trig. #	Mirror #	diff	Trig. #	Mirror #	diff
s 126				s 954		s 378		
c 612	c 216	c 396			c 396		s 738	
	s 162		s 594					c396
738 + 378 + 396 =	1512		594 + 954 + 396 =	1944		378 +738 +396=	1512	

- **378** feet is the half-base width of the Great pyramid.
- **612** feet is the apothem length.
- **3024** feet is the baseline perimeter (2 x 1512 = 3024)
- The ratio **378 : 612** is the cosine of the slope angle of the pyramid. i.e. **Slope angle = 51°.85548690 = 51° 51' 19.75"**

- Half value of (Pi) that was seemingly used in the construction of the Great Pyramid is given by the ratio **594 : 378**.
  Hence, Pyramid Pi $= 2 \times (594 / 378) = \pi_p = 3.142857 = 22/7$

- Pyramid height $= \sqrt{((612 \text{ ft})^2 - (378 \text{ ft})^2)} = 481.31$ ft. $\cong$ **481** ft.
  also,
  Pyramid height $= \sqrt{(612 \text{ ft} \times 378 \text{ ft})} = 480.97$ ft. $\cong$ **481** ft.

  Pyramid height $= (2 \times 378 \text{ ft})^2/(2 \times 594 \text{ ft}) = 481.09$ ft $\cong$ **481** ft.

- The square of the ratio of the pyramid's tabular apothem length to the calculated design height provides a close approximation to the Golden ratio, with a 0.05 % error. i.e.

  Phi $(\varphi) \cong$ **(612 ft : 481 ft)^2** $= (1.272349272)^2 = $ **1.6189**

- If the ascending alpha numbers from the three *Alpha-numeric Color Groups* and their Alphanumeric values are displayed as ratios, then summed, a value twice Pyramid Pi is obtained.
  i.e. **(126:144) + (459:144) + (378:170) = 6.2860** $\cong$ **$(2\pi_p)$**

  Twice Pyramid Pi $=$ **44/7 = 6.2857** (0.005 % diff.)

- It is interesting to note that the product of the **126** (blue color group) column sums has a value that reveals what Tenuta might call *"A Secret of Nine"* (from the title of his book) i.e.

  (126 × 144 × 18) = 326,592 = 0.9 × 9! = Nine Tenths of Nine Factorial.

The Face to Base Golden Ratio of the Great Pyramid And the Seventh Part of a Million Square Feet.

The Great Pyramid of Giza is the remaining *seventh* part of what has been referred to as the *Seven-Wonders of the World,* and it is extremely interesting to discover that a *seventh* part of a *million* square feet represents one quarter of its base area.

The following presentation will elucidate the exactness of the external measurements of W.M.F.Petrie[1] and reveal the extreme accuracy of the numerical comparisons taken from the column and row sums of the Glimmer-Tables.

- Division of a ***million square feet*** by ***seven*** forms a recursion number that is one-quarter of the total base area of the Great Pyramid i.e.
  Recursion number = $(1 \times 10^6$ ft$^2) / 7 = 142857.142857\cdots$ ft^2.

- The square root of the ***recursion number*** is the value of the Great Pyramid's baseline half-width. i.e.
  Half-width = $\sqrt{(142857.142857 \text{ ft}^2)} = 377.964$ ft $\cong$ *378 ft*.
  Baseline Width = 2 x 377.964 ft. = 755.929 ft. $\cong$ *756 ft*.

W.M.F. Petrie's[1] mean value for the width of the Great Pyramid is 9068.8 ± 0.5 inch = 755.73 ± 0.04 ft. This is an error in the recursion number resultant baseline width of ~ 0.03%.

- Multiplication of the Great Pyramid's half-width by the Golden ratio (φ) yields the pyramid's apothem value. i.e. Apothem = ($\sqrt{\text{Recursion \#}}$)($\varphi$) = 611.559 ft. $\cong$ *612* ft.

---

[1] W. M. F. Petrie – *"The Pyramids and Temples of Gizeh"* – Chapter 6. – Section 25.

- The area of one face of the Great pyramid is the half-width times the apothem value, thus the total surface area is,

  Area of Four Faces $= 4$(Half width)(Apothem)

  $\qquad\qquad\qquad = 4(\sqrt{\text{Recursion }\#})(\sqrt{\text{Recursion }\#})(\varphi)$

  $\qquad\qquad\qquad = 4(\text{Recursion }\#)(\varphi)$

  $\qquad\qquad\qquad = 4(142857.142857\text{ft}^2)(1.618034)$

  Area of Four Faces $\equiv (\varphi)(\text{Base Area}) = 924{,}590.85\ \text{ft}^2$

  $\therefore$ **The** face to base ratio of the Great Pyramid is the Golden Ratio Phi $(\varphi) = 1.618034$

- An approximate value for Phi (φ) is found with the square of the ratio of major sums from Table-1, and has been termed *Pyramid Phi* (φ_{p1}) by the author. $\sqrt{\varphi_{p1}} = \textbf{\textit{756/594}} = 1.2727\cdots$

  $\varphi_{p1} = (\textbf{1.272727}\cdots)^2 = \textbf{1.6198}$ 0.11% error   True value $(\varphi) = 1.6180$

- When the stated apothem and half-width values from Table-1 are used in a ratio, a more accurate value for Phi is obtained.

  $\varphi_{p2} = (\textbf{\textit{612}}\textbf{ ft} / \textbf{\textit{378 ft}}) = \textbf{1.6190}$ 0.06% error

---

- The "Great Pyramid Pi value" is a ratio of tabular sums from Table-1 and is termed the *Pyramid Pi* (π_p) by the author.

  $\pi_p = \textbf{\textit{1188 / 378}} = \textbf{22/7}$ 0.04% error

---

- When the *recursion number* is multiplied by the true **Phi** (φ) value, an approximately correct value for the square of the pyramid's design height is obtained.

  $(142857.142857\text{ ft}^2) \cdot (1.618033989) = 231{,}147.7128\text{ ft}^2$

  Height of G.P $= \text{sq.rt.}(\ 231{,}147.7128\text{ ft}^2) = 480.78\text{ ft.} \cong \textbf{481}\textbf{ ft.}$

Additionally, using Table-1 summation data.

- Height GP = Perimeter / $2\pi_p$ ≡ Base Area / *Total Sum* Table-1.

  Height = *3024 ft./(44/7)* ≡ *(756 ft)^2/1188 ft* = 481.1 ft. ≅ **481** ft.

Note: The whole numbers shown in bold face italics are numbers that can be found in the column and row sums, and certain area sums of the Glimmer-Tables . All have trigonometric values that can be expressed in terms of Phi according to the *Phi function selection rule.*

- ### *Slope Angle from Recursion Value is Phi (φ) Dependent*

  Cosine of baseline slope angle (β) = Half-width /Apothem

  Cosine (β) = ($\sqrt{}$Recur. #) / ($\sqrt{}$Recur. #)(Phi φ) = 1/ Phi φ

  Cosine (β) = 1/1.618033989 = 0.618033989

  Slope (β_1) = 51°.82729238 = 51° 49′ 38.25″ ≅ 51° 50′

- ### *Slope Angle from Tabular Values*
  Cosine of baseline slope angle (β) = Half-width / Apothem

  Cosine (β) = Half-width / Apothem = *378* ft. / *612* ft.

  Cosine (β) = 0.617647059

  Slope (β_2) = 51°.85548690 = 51° 51′ 19.75″ ≅ 51° 51′

- ## *Great Pyramid Slope Angle from Euler's Number (e)*

The value for (β), the base line slope angle of the Great Pyramid, is *independent* of any linear pyramid measurement, and has been shown[9] to be dependent only on the *Euler number2.71828.* i.e.

Ratio $\beta/\theta = e/2$ for the Great Pyramid right triangle.

$\beta \div (90° - \beta) = 2.71828 \div 2$

$2\beta = 2.71828 (90° - \beta)$

$2\beta = 2.71828 \times 90° - 2.71828 \times \beta$

$2\beta + 2.71828\,\beta = 244.6452$

$\beta = 244.6452 \div 4.71828 = 51°.8505$

$\beta = 51$ deg 51 min 01.8 sec $\cong$ 51 deg 51 min

$(\text{Tan } \beta)^2 = (\text{Tan } 51°.8505)^2 = (1.273081)^2 = 1.621 \cong$ **Phi (φ)**

0.17 % error

True value Phi(φ) = 1.6180

Compare to the Phi values obtained from Table-1 ratios.

$\varphi_{p1} = (756 / 594)^2 = 1.6198$ 0.11% error

$\varphi_{p2} = (612 / 378) = 1.6190$ 0.06% error

---

[9] Reference: *"The Euler Number and the Great Pyramid"* by Rick Howard–August 2002   http://www.gizapyramid.com/research article.

All of the values for the outside measure of the Great Pyramid determined here by using the quarter base area recursion value, 142857.142857 sq.ft., the true value of Phi, and the Euler number are all within the mean value range as measured by W.M.F. Petrie in his historical book, *The Pyramids and Temples of Gizeh* – Chap.6.– Sect. 25.

---

1. If (n) is an integer divisible by 9, and (n) ÷ 360 contains one decimal place, (i.e., .1, .2, .3, .4, .6, .7, .8, .9), excluding (.0 & .5), then the Cosine (n) can be expressed as a function of Phi. ---- i.e. cosine (n) = $f(\varphi)$. If (n) ÷ 360 ends with (.0 or .5), then Cosine (n) = ± 1.

2. If (n) is an integer divisible by 9, and (n) ÷ 360 contains two decimal places that are an odd multiple of (.05), (i.e., .05, .15, .35, .45, .55, .65, .85, .95) excluding (.25 & .75), then the Sine (n) can be expressed as a function of Phi. i.e. sine (n) = $f(\varphi)$. If (n) ÷ 360 ends with (.25 or .75), then Sine (n) = ± 1.

3. The numbers (n) that end with a 4 or 6 have a trig. function of ± φ/2. The numbers (n) that end with a 2 or 8 have a trig. function of ± 1/(2φ).

Figure-8   Trigonometric Phi(φ) Function Identity Rule

Summary

Glimmer Tables

The process of number reduction by digit addition called distillation, when applied to the Fibonacci series, led to the development of the first of several "Glimmer" tables. The column, row and numerically suggestive area summations of the parent table surprisingly provided numbers that are the exact external dimensions of the Great pyramid, as measured in feet. Other numbers that occur continually in the tables represent physical earth measurements of things such as size, surface area, equatorial rotation speed, angular velocity and rotational energy. The units of measure seemingly are those from ancient metrology, which apparently was the system the British later adapted for use in the formation of the Imperial system of measure.

These findings and others developed from the tables, along with the patterns of symmetry and mirror imagery present in them, reinforce the argument that similar such tables must have existed in the distant past.

One table of a different nature, Table-5, is an alpha-numeric chart with nine rows of letters and double letters placed in alphabetical order, and whose distilled values increase from one to nine proceeding downward. The notes provided in several rows are a mystery as to their *raison d'être,* but are there because of their mathematical existence and exactness to some of the essential numbers in this treatise. i.e.

- The numerical values of the letters B, K & T when multiplied have a product that is the baseline width of the Great pyramid if measured in Royal cubits. i.e. $2 \cdot 11 \cdot 20 = \mathbf{440}$ cubit
- The numerical values of the letters C, L & U in the next row have a product that, if measured in feet, is the base width of the Great pyramid i.e. $3 \cdot 12 \cdot 21 = \mathbf{756}$ feet.

- The 6th row product of the letters F,O, & X = 6•15•24 = **2160** is the mean value of the moon's diameter measured in s.miles.
- The 7th row value for the letters G, P, & Y when multiplied is 7•16•25 = 2800, which surprisingly is a base ten harmonic of the height of the Great pyramid as measured in cubits. i.e. Height = **280** cubits.

It is interesting that the letters G, P, & Y could be an acronym for the **G**reat **P**yramid's **Y**-axis.

The Wheel of Phi (φ)

The cyclic rhythms of our ancient universe were surely the basis of man's concept of *time*. The orbital rotations of the heavenly bodies that he observed required him to conceive a means to measure such movement. The "amount of movement" was most certainly based on the daily rising and setting of the sun and the seasonal changes that occurred before the long cycle was repeated. The choice of the number 360 as the number of *"degrees"* for this annual movement was quite possibly a result of the number of days they counted in their early measurement of annual time. i.e. The *"daily egress"* of the rising sun.

The choice of 360 degrees per cycle has been accepted and used since the first early application in antiquity and remains in use today. Division of a circle into twenty equal angles as shown by the twenty spoke Phi (φ) wheel of Figure-6 illustrates why ancient artists and architects might have chosen these angular numbers for use as linear measurements. It is apparently because so many of these same numbers or fractals thereof are found to be prevalent in nature, and notably the trigonometric value of these numbers can be expressed in terms of Phi(φ) according to the Trigonometric Phi(φ) Function Selection Rule. The exceptions to the rule are the cardinal compass points (90°-180°- 270°-360°), which are precisely the angular directions of each pyramid face.

The essay on Leonardo da Vinci's *Vitruvian Man* shows an aspect of *Nature's numbers* that heretofore seems not to have been generally known. The angles chosen by da Vinci in his drawing of the open leg stance with arms raise overhead are angles that have trigonometric values that can be expressed in terms of Phi(φ). The angular placement of the arms over the head at 144 degrees divides the circle into two parts. This is a ratio of 144:216

Did da Vinci select this angular ratio 144:216 for placement of the arms of the "human male" in the circle, or was it just an unavoidable fact that was by nature the only possible choice?
It is the authors belief that the angular perspective presented here, concerning the use of angles whose trigonometric values can be expressed in terms of Phi(φ), offers further evidence of nature's influence on the great works of man.

Some of the angles from the Phi (φ) wheel may also be found in the biology of plants and flowers in the wild, and a common springtime weed called Daisy Fleabane is examined herein to provides the reader a look at its five leaf angular rotation patterns.

The *Glimmer Tables* and the *Wheels of Phi(φ)* together form a root source for *Nature's* numbers from which present day artists and architects might select angles and numbers that are thought to infuse aesthetical value into their designs and works of art.

~ ~ ~ ~

PART 3

~~~

Author Side Notes

~~~

Glossary

~~~

References

Joseph
Turbeville

Author's Side Notes

Side Note - 1

An early "nudge" in the development of these numerical tables came while reading some of the papers written by Charles Johnson[5] in his *Earth/matriX, Science in Ancient Artwork* series.

In his writing on historically significant numbers, he states that, "not only do they correspond to the simplicity of geometrical designs in ancient artwork, but the numbers correspond to phenomena occurring in nature". It was his mention of the word "nature" that suggested the mathematical growth series of Fibonacci, a subject that I had been exploring for some months. It was at this moment that I felt compelled to include the search for "historically significant numbers" as a major part of my study.

The creation of the first numerical table grew from an idea to reduce the Fibonacci series[*] to single digits by the process of numerical distillation just to see what the results might reveal. The numbers in the series, when distilled, were found to have an upper limit of 24 digits, at which point a continuation of the distillation process simply reproduces the original ordered digits.

The thought of creating nine rows of 24 digits by a process of multiplication and distillation was the genesis of all the tables in this report. The discovery of patterns of symmetry and mirror imagery, in addition to the discovery that certain historically significant numbers seemed to repeatedly occur as sums in the different geometric patterns, were all exciting discoveries which seemed to urge me on.

---

[*] See Table-6   First twenty-four Fibonacci numbers.

95

For many years, my intuition has at times come to my aid and provided the insight needed to solve some problem I may have been struggling with. Often coincidental happenings, with no apparent causal relation to a problem, might somehow seem meaningful and move me in the direction of a solution. Such happenings are called synchronistic events.

Early on in the development of these tables, three seemingly unconnected events happened that have had a profound effect on this work. All occurred on the same day and I now refer to them as a *triple-synchronistic event.*

Around 7:45 a.m. on May 26, 1998, a friend in Florida sent me an e-mail which included a dirty little sound bite that was evidently an imitation of Donald Duck. Around 1:00 p.m. my older son sent an e-mail and signed off with something that reminded me of one of the Disney characters, i.e. (that's aaaaaaaaaaaall folks). I think it was Porkey the Pig. That reminded me of the earlier e-mail I had received. Again, that evening around 7:45 p.m., while watching Jeopardy on TV, one of the statements concerned the answer (as I recall) who was Daffy Duck? At the very moment I heard the answer, I was holding in my hand a printout of my son's e-mail. Then it hit me. A *triple synchronistic event!* I suddenly realized that these three unconnected events were somehow trying to tell me something.

Later the same evening while searching the Internet for additional information on synchronicity, I ran across an article I had never before seen titled *"The 22 Enigma"*, by Terry Alden. After reading the article, I knew we had a lot in common and felt as though I had somehow been guided to his website. Later correspondence with Terry has proven this to be one of the connections I was to make and it has had a definite positive effect on my research. We made a connection because the number **22** has always been a number that repeatedly shows up in my life. I am for a fact, what might be called a *twenty-two* person.

In early spring of 1999, I received an unusual e-mail from my friend in Florida again. So unusual in fact, that as soon as I read the first three lines, I was immediately reminded of the triple synchronistic event which he initiated some months earlier. Here, in part, is exactly what he said: *"Well, Joe I guess spring is here as I spent the entire day raking oak leaves from the front and side yard and I think I picked up one million nine hundred thousand eight hundred and thirty-five of the sob's, my back is killing me tonight. ·········"*

Seeing the numerical values **1,900,800** and **35** just as he had written them really got my attention. The larger number was one of the most interesting numbers from antiquity that I had uncovered in my research. The thing that really ties it to the *triple synchronistic event* is the fact that **1,900,800** is precisely the number of seconds in **22** days, and **22** is the subject of Terry Alden's paper referenced above. Also, the number **35** just happens to be the alphanumeric value of my initials (JET). To me, this was truly a positive synchronistic happening that taught me to relax and "go with the flow". I try now, to always be receptive and aware of any little "nudge" that life may give me.

Figure-3A on p.98 lists the diameters of the planets and the Earth's moon in a manner similar to Figure-3 on page 28, with the exception that the values are in *statute miles* and they too have been finitely adjusted so that all can be distilled to nine. The average error is again less than two tenths of one percent. The reader should find it interesting that several of the numerical values for the diameters are identical to, or a base ten multiple of those found in the tables. e.g.. **864, 3024, 756, 2160.**

```
DIAMETER                                    %ERROR

SUN  864,000 s.mile   x  1.609344 km./s.mile   =  1,390,473 km.
2400 x 1,900,800 ft.     True value 1,391,900 km ⇒ Error 0.10%

MERCURY  3024 s.mile  x  1.609344 km./s.mile   =   4866.656 km.
8.4 x 1,900,800 ft.      True value  4,866 km.  ⇒ Error 0.01%

VENUS  7560 s.mile    x  1.609344 km./s.mile   =  12,166.641km.
21 x 1,900,800 ft.       True value  12,106 km. ⇒ Error 0.50%

EARTH   7920 s.mile   x  1.609344 km./s.mile   =  12,746.005 km.
22 x 1,900,800 ft.       True value  12,743 km. ⇒ Error 0.02%

MOON   2160 s.mile    x  1.609344 km./s.mile   =   3,476.183 km.
6 x 1,900,800 ft.        True value  3,476 km  ⇒ Error 0.01%.

MARS   4212 s.mile    x  1.609344 km./s.mile   =   6,778.557 km.
11.7 x 1,900,800 ft.     True value  6760 km.  ⇒ Error 0.27%

JUPITER  86760 s.mile x  1.609344 km./s.mile   =  139,626.685 km.
240 x 1,900,800 ft       True value 139,516 km. ⇒ Error 0.08%

SATURN  72360 s.mile  x  1.609344 km./s.mile   =  116452.132 km.
201 x 1,900,800 ft.      True value 116,438 km. ⇒ Error 0.01%

URANUS  29160 s.mile  x  1.609344 km./s.mile   =   46928.47 km.
81 x 1,900,800 ft.       True value  46940 km. ⇒ Error 0.02%

NEPTUNE 28188 s.mile  x  1.609344 km./s.mile   =  45,364.189 km.
78.3 x 1,900,800 ft.     True value  45432 km. ⇒ Error 0.15%

PLUTO  7092 s.mile    x  1.609344 km./s.mile   =  11,413.467 km.
19.7 x 1,900,800 ft.     True value  11400 km. ⇒ Error 0.12%
```

Figure-3A Planets and Major Moon Diameters in S. Miles

Side Note -3

Having recently gone through several days of writers block brought on by a severe case of spring fever, I was spending time over my morning coffee composing an e-mail to my daughter to thank her for the pictures she had just sent us of flowers around her new home. As I finished my note to her, and at the very moment I was about to hit the send button, we had a momentary power outage! (In this rural area it happens quite often. Needless to say it is always very annoying). No longer in the mood to try and retype my letter, and disgusted with my recent writing skills in general, I decided to look through my file cabinet for a particular photograph my wife had asked me to find. While in the process, I came across an architectural drawing that my daughter had sent me some time back when she was developing a design plan for her thesis. *Then it hit me!*

Her plan-view drawing incorporated two Golden rectangles placed at an obtuse angle to form the main design structure. This image and the angles she presented there reminded me of the Vitruvian man and the angular values Leonardo da Vinci had used in his drawing. Seeing my daughters angular, mirror image values, jolted me back to what I had been trying to write about when I began to suffer with the "writers disease." This sort of mental *"nudge"* is the result of what I refer to as a synchronistic event.

I immediately began to mull over my newly evoked ideas. I applied the trigonometric Phi function selection rule to her major design angles and was extremely surprised and excited to see that they also could be expressed in terms of Phi (φ). I decided to inform her of my discovery, as I knew she would be thrilled with this additional disclosure of Phi (φ) in her design angles. She had purposely developed her design using Golden rectangles with the intention of enhancing aesthetic value.

The use of the Golden ratio and Golden rectangles by architectural designers and artists has for centuries been thought to enhance the aesthetic value and beauty of their work.

The inclusion of angles and mirror image angular pairs, whose trigonometric values can be expressed in terms of Phi (φ), must surely evoke similar subliminal feelings and enhance aesthetic values. Practitioners of all the arts and sciences should be knowledgeable of, and perhaps make use of such natural angles in their designs whenever possible.

A cutaway section of the plan view of my daughters design shows the mirror image angles 117°+ 117° = 234° and also angles 63° + 63° = 126°, both of which satisfy the trigonometric Phi (φ) function selection rule discussed in Part 2.

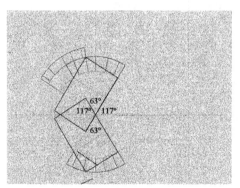

The sine of 234° = - φ/2 and the sine of 126° = + φ/2

This a fine example of how the choice of an angle or a mirror image pair of angles in a work of art or an architectural design might enhance the aesthetic value and beauty of the work, if only in a subliminal manner.

Side Note - 4

This note is one I was hesitant to include because it has little to do with the subject matter of the book, but everything to do with its publication. It concerns an extremely unusual, I should say almost impossible set of synchronistic happenings, and I only decided to write about it now because I feel it is the reason I was able to finish writing this expanded edition of the earlier book. The events really "nudged me" to get it all together and publish. Surely, there is a significant reason for this "happening"

As you review the chronology that follows, be reminded that the number **22** that pervades here was the catalyst for my first book, and was discussed in Side Note-2.

2/20/2002 5:37 PM (Note the **2**'s in time e-mails are sent)

Having learned in the winter newsletter message from Bruce Batchelor that Trafford had opened a branch office somewhere in the New Bern, NC came as a pleasant surprise to me, for it is the town where I was born. Fond memories immediately flooded my mind. I recalled spending many fun summers there at my grandparent's home on South Front Street as a youngster. What a coincidence, I was compelled to e-mail Carol and Mike Reed and introduce myself. I told them about the house I was born in being very close to the point where the Neuse and Trent rivers come together and how I learned to swim and sail small boats there.

2/22/2002 12:23 PM (Note the **2**'s in time e-mails are sent)

Mike Reed responds with a pleasant e-mail describing how the old town has changed, that there is a Sheridan Hotel and an excellent marina located there by the Trent River now. He describes the big bridge that has since been built there that rises 65 feet in the air to allow large sail boats to pass under.

2/6/2003

One year has passed since the exchange of e-mails with Mike Reed at Trafford in New Bern. While reading through the Trafford website on this date I discovered that the New Bern office had moved to an address on South Front Street. To be more precise, the address was **8 South Front Street**. Directly across the street from where I was born. I was amazed at this coincidence. This was truly a synchronistic event for me. I was so moved that I e-mailed Bruce Batchelor, Trafford's CEO and related the events of the earlier e-mail exchanges with Mike to him. He seemed to understand how my involvement with Trafford Publishers several years before their move to the almost exact location of my birth could create for me this very meaningful experience, a synchronistic event.

Photograph taken while on vacation in 1968 visiting my birthplace on South Front St. in New Bern, the former home of my grand parents Joseph and Kate Lucas. The place was dismantled sometime later to make way for a new bridge over the Trent River.

Glossary

Earth Data	Present day value	Normally Used

Mean radius   = 3958.95 s.mile ± 0.25 s.mile   3960 s.mile
Equatorial radius = 3963.44 s.mile ± 0.25 s.mile
Polar radius   = 3950.00 s.mile ± 0.25 s.mile

Statute mile   =   5280.00 feet   =   1609.344   meter
Nautical mile   =   6076.1155 feet =   1852.000   meter
Khufu mile[*]   =   6048.00 feet   =   1843.430   meter

Equatorial circumference based on Tabular dimensions of G.P.

English measure ⇒ 24742 s.mi.   (True value = 24902 s.mi)
Metric measure ⇒ 39818 kilometer (True value = 40076 km)
"Khufu" measure ⇒ 21600 kfu.mi. (True value = 21740 kfu.mi)

Polar circumference based on 481 ft. Height of G.P. & Pi=22/7

English measure ⇒   24737 s.mile (True value = 24860 s.mi.)
Metric measure ⇒   39811 kilometer (True value = 40008 km.)
"Khufu" measure ⇒ 21596 kfu.mi. (True value = 21704 kfu.mi.)

Symbols

± plus or minus	(÷ or / divide by)	*! factorial*
+ plus or add	= equal to	≡ identical with
− Subtract or minus	≈ or ≅ approx. equals	: ratio of
• or **x** multiply by	∴ Therefore	Normal to = ⊥

---

[*]   Assumed "inherent value" based on total sum of Table-2.

References

1   Petrie, WM - "*The Pyramids and Temples of Gizeh*" - Chapter 20 -
    Section 136 - Published in London 1883.   see text p.40

2   Tenuta,G. 1999. *The Secret Of Nine,* self published by author, E-mail,
    CodeUFO@aol.com   see text p.19

3   Tompkin, P. 1971. *Secrets of the Great Pyramid,* published by
    Galahad Books 1977. A division of Budget Book Service Inc. 386 Park
    Avenue South, New York, NY 10016. Published by arrangement with
    Harper Collins Publishers, 10 East 53rd Street, New York,   see text p.19

4   Furia, J. 1998. *"Evidence Of a Mathematically Created Solar System".*
    Geomusic Homepage, James Furia@aol.com   see text p.23

5   Johnson, C. 1996. Earth/matriX Science in Ancient Artwork, Series
    No.77. *The Great Pyramid.* New Orleans, Louisiana
    http://www.earthmatrix.com/series 77/pyramid.htm   see text p.23

6   Michell, J - "*The New View Over Atlantis*"- pp 132-133 - Thames and Hudson
    Ltd. - London, 1993   see text pp.25, 27

7   M. Chatelain – "*Our Cosmic Ancestors*"- p.34 - ISBN 0-929686-00-4
    Temple Golden Publication- Sedona, Arizona -   see text p.35

8   *McClain, E.G.– "The Pythagorean Plato: Prelude to the Song Itself"- p.23*
    ISBN 0-89254-010-9 –1984 - Publisher Nicolas-Hays, Inc.- York Beach,
    Maine-03910   see text p.73

9   *Howard, R. - "The Euler Number and the Great Pyramid"* - August 2002
    Research Article - http://www.gizapyramid.com/research   see text p.88

10 Lemesurier, P. -*"Decoding the Great Pyramid - ISBN 0-7607-1962-4*
    Published by 2000 Barnes and Nobel Books.

11 CRC "*Handbook of Chemistry and Physics*", 1981- 82. CRC Press, Inc.
    Boca Raton, Florida.

12 Knott, R -- http://www.mcs.surrey.ac.uk/Personal/R.Knott/Fibonacci
                      /fibnat.html#plants   see text p.77

PART 4

~~~

All Tables

~~~

Joseph
Turbeville

Parent Table from "A Glimmer of Light from the Eye of a Giant"

216 Cells arranged in **9** repeating rows of **24** distilled Fibonacci numbers in **24** columns, where each successive row is obtained by multiplying the row number by each respective top row factor and distilling. (This limited grid pattern will duplicate itself if continued in either the X or Y direction).

1	2	3	4	5	6	7	8	9	10	11	12	13	14	15	16	17	18	19	20	21	22	23	24	Row Sums ▽
1	2	3	5	8	4	3	7	1	8	9	8	8	7	6	4	1	5	6	2	8	1	9	1	117
2	4	6	1	7	8	6	5	2	7	9	7	7	5	3	8	2	1	3	4	7	2	9	2	117
3	6	9	6	6	3	9	3	3	6	9	6	6	3	9	3	3	6	9	6	6	3	9	3	135
4	8	3	2	5	7	3	1	4	5	9	5	5	1	6	7	4	2	6	8	5	4	9	4	117
5	1	6	7	4	2	6	8	5	4	9	4	4	8	3	2	5	7	3	1	4	5	9	5	117
6	3	9	3	3	6	9	6	6	3	9	3	3	6	9	6	6	3	9	3	3	6	9	6	135
7	5	3	8	2	1	3	4	7	2	9	2	2	4	6	1	7	8	6	5	2	7	9	7	117
8	7	6	4	1	5	6	2	8	1	9	1	1	2	3	5	8	4	3	7	1	8	9	8	117
9	9	9	9	9	9	9	9	9	9	9	9	9	9	9	9	9	9	9	9	9	9	9	9	216
45	45	54	45	45	45	54	45	45	45	81	45	45	45	54	45	45	45	54	45	45	45	81	45	1188

| 189 | | | | 189 | | | | 216 | | | | 189 | | | | 189 | | | | 216 | | | |
| 594 | | | | | | | | | | | | 594 | | | | | | | | | | | |

TABLE-1

Parent Table Notes

- Baseline Width of the Great Pyramid $= 1188 - 2(216) = 4(189) = 756$ ft.

- The total sum minus the sum of all 9's $= 1188 - 9(48) = 1188 - 432 = 756$ ft.

- The digit sum of the two central squares (bold white digits) $= 216 \Rightarrow 216$
  is the number of all gray cells.

- The sum of the perimeter digits (bold black) $= 612 \Rightarrow$ Apothem $= 612$ ft.

- The sum of the forty border gray nines (gray background) $= 9(40) = 360$
  360 degrees in a circle. --- **360** A yearly day count in antiquity.

- Great Pyramid scale size to the Earth $= (1 : 43200) \Rightarrow 2(216) = 432$
  432 is a base ten factor of 43200.

- Great Pyramid Pi Factor (π_p) $= 594 \div 189 = 22/7 = 3.142857$

- Great Pyramid Height $=$ Base Area Circumference $\div 2\pi_p$
  $(4 \times 756$ ft$) \div 2(22/7) = 481.1$ feet

- Great Pyramid Height $=$ Base Area $\div$ Total Sum of Table-1
  $(756$ ft$)^2 \div 1188$ ft. $= 481.1$ feet

- Time note : $(189,216,000$ sec.$) \div (86,400$ sec/day$) \div (365$ day/yr.$) \equiv 6$ year.

The Right Half Bold Digits of Table-1 when positioned directly above the Left Half Bold Digits form a **MIRROR IMAGE** which makes the numerical equality of the two sets immediately obvious to the reader. Column sums are shown at the bottom of the table.

Right Half Mirror Image Portion of Table-1 placed above the Left Half. Mirror Portion

TABLE-1B

8	8	7	6	4	1	5	6	2	8	1
7	7	5	3	8	2	1	3	4	7	2
6	6	3	9	3	3	6	9	6	6	3
5	5	1	6	7	4	2	6	8	5	4
4	4	8	3	2	5	7	3	1	4	5
3	3	6	9	6	6	3	9	3	3	6
2	2	4	6	1	7	8	6	5	2	7
1	1	2	3	5	8	4	3	7	1	8
1	1	2	3	5	8	4	3	7	1	8
2	2	4	6	1	7	8	6	5	2	7
3	3	6	9	6	6	3	9	3	3	6
4	4	8	3	2	5	7	3	1	4	5
5	5	1	6	7	4	2	6	8	5	4
6	6	3	9	3	3	6	9	6	6	3
7	7	5	3	8	2	1	3	4	7	2
8	8	7	6	4	1	5	6	2	8	1
1 72	72	72	90	72	72	72	90	72	72	72
216			90	216			90	216		

Total Sum = **828** Sum of Central Rectangles (White digits) = **216**
Half Sum = **414** Sum of Perimeters (Bold Black digits) = **612**
- The apothem length of the Great Pyramid is **612** feet

By superimposing each half of the Mirror Image Table directly over one another and summing the overlaying 88 cells the following number distribution is obtained that also has mirror image properties.

9	9	9	9	9	9	9	9	9	9	9	198
9	9	9	9	9	9	9	9	9	9	9	
9	9	9	18	9	9	9	18	9	9	9	216
9	9	9	9	9	9	9	9	9	9	9	
9	9	9	9	9	9	9	9	9	9	9	216
9	9	9	18	9	9	9	18	9	9	9	
9	9	9	9	9	9	9	9	9	9	9	198
9	9	9	9	9	9	9	9	9	9	9	
216			90		216		90		216		

Dark Gray Area	+	Light Gray Area	=	Table Sum
16 x 9 = 144		68 x 9 = 612		84 x 9 = 756
+ 4 x 18 = 72				+ 4 x 18 = 72
216	+	612 (Mirror Image Numbers)		= 828

- The Width of the Great Pyramid at Giza is **756** feet .
- The Apothem of the Great Pyramid is **612** feet.

TABLE-1C

112

9	9	9	9	9	9	9	9	9	9	9	9	9
9	**1**	**1**	**2**	**3**	**5**	**8**	**4**	**3**	**7**	**1**	**8**	9
9	**2**	**2**	**4**	**6**	**1**	**7**	**8**	**6**	**5**	**2**	**7**	9
9	**3**	**3**	**6**	**9**	**6**	**6**	**3**	**9**	**3**	**3**	**6**	9
9	**4**	**4**	**8**	**3**	**2**	**5**	**7**	**3**	**1**	**4**	**5**	9
9	**5**	**5**	**1**	**6**	**7**	**4**	**2**	**6**	**8**	**5**	**4**	9
9	**6**	**6**	**3**	**9**	**3**	**3**	**6**	**9**	**6**	**6**	**3**	9
9	**7**	**7**	**5**	**3**	**8**	**2**	**1**	**3**	**4**	**7**	**2**	9
9	**8**	**8**	**7**	**6**	**4**	**1**	**5**	**6**	**2**	**8**	**1**	9
9	9	9	9	9	9	9	9	9	9	9	9	9

Table-1D   Modified Half View

Cell Area	Cell Count	Value
• Central Gray Cube	(20 cells)	Digit Sum = 108
• White Perimeter (Blk.Digits)	(22 cells)	Digit Sum = 99
• Gray Perimeter (Blk.Digits)	(30 cells)	Digit Sum = 135
• White 9's in Outer Perimeter	(42 cells)	Digit Sum = **378**
• All Bold Black Digits	(88 cells)	Digit Sum = 414
Minus 4 Large Corner 9's	– (4 cells)	Digit Sum = -36
	84 cells)	**378**

- Half the baseline width of the Great Pyramid is **378** feet.

When the two halves of Table-1 are placed one over the other and the overlaying digits are added, the following numerical pattern is formed.

Row Digit Sum ∇

9	9	9	9	9	9	9	9	9	9	9	18	117
9	9	9	9	9	9	9	9	9	9	9	18	117
9	9	9	18	9	9	9	18	9	9	9	18	135
9	9	9	9	9	9	9	9	9	9	9	18	117
9	9	9	9	9	9	9	9	9	9	9	18	117
9	9	9	18	9	9	9	18	9	9	9	18	135
9	9	9	9	9	9	9	9	9	9	9	18	117
9	9	9	9	9	9	9	9	9	9	9	18	117
18	18	18	18	18	18	18	18	18	18	18	18	216
90	90	90	108	90	90	90	108	90	90	90	162	**1188**
378				378				432				
756												

TABLE-1 OVL

- Base Width of Great Pyramid (Khufu) = **756** feet
- Apothem Length of Khufu = **612** feet = Sum of 68 Light Gray 9's
- True compass heading of Khufu's Northernmost face = **360°**
    Sum of the 20 white on gray border 18's = **360**
- Sum of all 18's = **432** ⇒ Great pyramid to Earth scale (**1 : 43200**)
- Khufu's Pi factor (π) = **(1188 / 378)** = **22/7** = **3.142857**
- Khufu's Height = $((612)^2 - (378)^2)^{1/2}$ = **481.3** feet

114

108 Cells of 9 repeating rows of 24 distilled Fibonacci numbers divided into12 columns (2 digits per cell), in which each digit of each successive row is obtained by multiplying the row number by each respective top row factor and distilling. (This is a limited grid pattern that will repeat itself if continued.)

▽ DOUBLE DIGIT ROW SUM   SINGLE DIGIT ROW SUM ▽

603	11	23	58	43	71	89	88	76	41	56	28	19	117
603	22	46	17	86	52	79	77	53	82	13	47	29	117
621	33	69	66	39	33	69	66	39	33	69	66	39	135
603	44	83	25	73	14	59	55	16	74	26	85	49	117
603	55	16	74	26	85	49	44	83	25	73	14	59	117
621	66	39	33	69	66	39	33	69	66	39	33	69	135
603	77	53	82	13	47	29	22	46	17	86	52	79	117
603	88	76	41	56	28	19	11	23	58	43	71	89	117
1188	99	99	99	99	99	99	99	99	99	99	99	99	216
6048	495	504	495	504	495	531	495	504	495	504	495	531	**1188**
	999		999		1026		999		999		1026		
	3024						**3024**						

Table-2

- The sum of either central 12 cell rectangle (white digits) is **612** ⇒ Apothem of Great Pyramid = **612** feet

- A length of **6048** feet is termed by the author as a *Khufu Mile.*

- The circumference of the Great Pyramid is (4 x **756** feet) = **3024** feet

The Bold numbers in the right half of Table-2 when positioned directly above the Bold numbers of the left half, form a MIRROR IMAGE, which makes the numerical equality of the two sets immediately obvious.

88	76	41	56	28	19
77	53	82	13	47	29
66	39	33	69	66	39
55	16	74	26	85	49
44	83	25	73	14	59
33	69	66	39	33	69
22	46	17	86	52	79
11	23	58	43	71	89
11	23	58	43	71	89
22	46	17	86	52	79
33	69	66	39	33	69
44	83	25	73	14	59
55	16	74	26	85	49
66	39	33	69	66	39
77	53	82	13	47	29
88	76	41	56	28	19
792	810	792	810	792	864

TABLE-2B

- Sum of either central 12 cell rectangle.   = **612**
  Apothem length of the Great Pyramid.   = **612** ft.
- Half-Sum of right side dark gray column.   = **432**
  Great Pyramid to Earth scale ratio. (**1 : 43200**)

116

When one half of the Mirror Image Table-2B is positioned directly atop the other and the overlaying numbers are summed, they create another Mirror Image pattern with the following interesting number distribution which is quite similar to Table-2 OVL, with the exception of the ninth row.

∇ Row Number Sum				Row Digit Sum ∇			
603	99	99	99	99	99	108	99
603	99	99	99	99	99	108	99
621	99	108	99	108	99	108	81
603	99	99	99	99	99	108	99
603	99	99	99	99	99	108	99
621	99	108	99	108	99	108	81
603	99	99	99	99	99	108	99
603	99	99	99	99	99	108	99
4860	792	810	792	810	792	864	756

TABLE-2C

- Base Width of the Great Pyramid (Khufu) = (2 x 378 ft) = **756** feet
- Apothem Length of Khufu = (4 x **99** + 2 x **108**) = **612** feet
- Khufu's Pi factor (π) = ((12 x 99) ÷ **378**) = 22/7 = 3.142857
- Khufu's Design Height = $((612)^2 - (378)^2)^{1/2}$ = **481.3 feet.**

117

Double-Digit plus Single Digit Summation by Column & Row of Distilled Fibonacci Numbers Presented in Table-2

▽ Black Digit Row Sum White Digit Row Sum Row Total ▽

Black Digit Row Sum													White Digit Row Sum	Row Total
347	13	28	71	50	79	106	104	89	46	67	38	29	373	720
361	26	56	25	100	59	95	91	61	92	17	58	40	359	720
375	39	84	78	51	39	84	78	51	39	84	78	51	381	756
353	52	94	32	83	19	73	65	23	85	34	98	62	367	720
367	65	23	85	34	98	62	52	94	32	83	19	73	353	720
381	78	51	39	84	78	51	39	84	78	51	39	84	375	756
359	91	61	92	17	58	40	26	56	25	100	59	95	361	720
373	104	89	46	67	38	29	13	28	71	50	79	106	347	720
702	117	117	117	117	117	117	117	117	117	117	117	117	702	1404
618	585	603	585	603	585	657	585	603	585	603	585	657	3618	7236
	1188		1188		1242		1188		1188		1242			

Table 2-D

- Great Pyramid baseline width = 756 feet.
- Corner line length ≅ 720 feet.

119

When the two halves of Table-2 are placed one over the other and overlaying numbers are added, the following numerical pattern
  is formed

∇	Row Number Sum				Row Digit Sum		∇
603	99	99	99	99	99	108	99
603	99	99	99	99	99	108	99
621	99	108	99	108	99	108	81
603	99	99	99	99	99	108	99
603	99	99	99	99	99	108	99
621	99	108	99	108	99	108	81
603	99	99	99	99	99	108	99
603	99	99	99	99	99	108	99
1188	198	198	198	198	198	198	sum 756
	6048						

TABLE-2 OVL

- Base Width of Great Pyramid - Khufu = **756** feet
- Apothem Length of Khufu = **612** feet = (4 x 99)+ (2 x 108)
- Perimeter of Great Pyramid = **3024** feet = **0.5** Khufu mile
- One Khufu Mile = **6048** feet (inherent tabular value)
- Khufu's Pi factor = (**(2 x 1188)** ÷ **756**) = **22/7 = 3.142857**

72 Cells arranged in **9** repeating rows of **24** distilled Fibonacci numbers divided into **8** columns (3 digits per cell), in which each digit of each successive row is obtained by multiplying the row number by each respective top row factor and distilling.

(This is a limited grid pattern that will repeat itself if continued in either the X or Y direction).

	TRIPLE- DIGIT ROW SUM ▽								SINGLE DIGIT ROW SUM ▽
4005	112	358	437	189	887	641	562	819	117
4005	224	617	865	279	775	382	134	729	117
4995	336	966	393	369	663	933	696	639	135
4005	448	325	731	459	551	674	268	549	117
4005	551	674	268	549	448	325	731	459	117
4995	663	933	696	639	336	966	393	369	135
4005	775	382	134	729	224	617	865	279	117
4005	887	641	562	819	112	358	437	189	117
7992	999	999	999	999	999	999	999	999	216
42012	4995	5895	5085	5031	4995	5895	5085	5031	1188

Table-3

- Triple-digit sum + Single-digit sum
      42012   +   1188   =   43200
  Great Pyramid to Earth Scale Ratio   =   (1 : 43200)

- <u>Angular rotation units:</u>   • <u>Time units:</u>

                                    21600 sec = **6** hour
21600 arc min = **360°** = **1 Rev**   **43200** sec = **12** hours
                                      43200 min = **30** days

The bold black numbers from the right half of Table-3 when positioned directly above the bold black numbers on the left form a MIRROR IMAGE which makes the numerical equality of the two sets immediately obvious.

887	**641**	**562**	**819**	2909
775	**382**	**134**	**729**	2020
663	**933**	**696**	**639**	2931
551	**674**	**268**	**549**	2042
448	**325**	**731**	**459**	1963
336	**966**	**393**	**369**	2064
224	**617**	**865**	**279**	1985
112	**358**	**437**	**189**	1096
112	358	437	189	1096
224	617	865	279	1985
336	966	393	369	2064
448	325	731	459	1963
551	674	268	549	2042
663	933	696	639	2931
775	382	134	729	2020
887	641	562	819	2909
7992	9792	8172	8064	

TABLE-3B

Sum Total $= (180 \times 189) = (45 \times 756) = 34020$

122

When the top eight rows from each half of Table-3 are superimposed and the overlaying numbers are added, the following Mirror Image pattern is formed.

∇	Row Number Sum			Row Digit Sum	∇
4005	999	999	999	1008	90
4005	999	999	999	1008	90
4995	999	1899	1089	1008	81
4005	999	999	999	1008	90
4005	999	999	999	1008	90
4995	999	1899	1089	1008	81
4005	999	999	999	1008	90
4005	999	999	999	1008	90
34020	7992	9792	8172	8064	702

TABLE-3C

Sum Total = (180 x 189) = (45 x 756) = 34020

Earth's "g" value at 45° ≅ $\sqrt[3]{(34020 - 702)}$ ≅ **32.178** ft./sec.^{2}

123

Triple-digit plus single-digit summation by column & row
of distilled Fibonacci numbers presented in Table-3

ROW TOTAL SUM - ∇

∇ - BLACK ROW SUM					WHITE DIGIT ROW SUM - ∇					ROW TOTAL SUM - ∇	
1148	116	374	451	207	910	652	575	837		2974	4122
2044	232	631	884	297	794	395	142	747		2078	4122
2130	348	987	408	387	678	948	717	657		3000	5130
2018	464	335	742	477	562	691	284	567		2104	4122
2104	562	691	284	567	464	335	742	477		2018	4122
3000	678	948	717	657	348	987	408	387		2130	5130
2078	794	395	142	747	232	631	884	297		2044	4122
2974	910	652	575	837	116	374	451	207		1148	4122
4104	1026	1026	1026	1026	1026	1026	1026	1026		4104	8208
21600	5130	6039	5229	5202	5130	6039	5229	5202		21600	43200

© 1999, Joseph Turbeville

TABLE-3D

- Equatorial Circumference in Khufu Miles = **21600** kfu.mi. = **24742** s. mile
- The scale size of the Great Pyramid to the Earth is (**1 : 43200**)

Angular rotation units	Time units	
21600 arc min = **360°** = **1 Rev**	**21600** sec = **6 hr**	**21600** min = **15 day**
43200 arc min = **720°** = **2 Rev**	**43200** sec = **12 hr**	**43200** min = **30 day**

When the two halves of Table 3 are superimposed and the overlaying numbers are added, the following numerical pattern is formed.

∇	Row Number Sum				Row Digit Sum	∇
4005	999	999	999	1008		90
4005	999	999	999	1008		90
4995	999	1899	1089	1008		81
4005	999	999	999	1008		90
4005	999	999	999	1008		90
4995	999	1899	1089	1008		81
4005	999	999	999	1008		90
4005	999	999	999	1008		90
7992	1998	1998	1998	1998		108
42012	9990	11790	10170	10062		810

TABLE-3 OVL

Table-3 digit sum plus OVL sum = 43200
i.e.   **1188 + 42012 = 43200**
Pyramid to Earth scale ratio   **(1 : 43200)**

126

A Partial Assembly of Numbers (Distillable to 9) Found in the Table Summations

The bottom row of numbers in the table below form a limited set of baseline numbers and the all white column numbers on the dark gray cells represent the first eight terms of the Fibonacci Series for each respective row. Column base multipliers (9, 18, 36, 72, 144 & 288) form 23 ascending columns of *duplacio* numbers, many of which are subjects of this work.

6	288	576	864	1152	1440	1728	2016	2304	2592	2880	3168	3456	3744	4032	4320	4608	4896	5184	5472	5760	6048	6336
5	144	288	432	576	720	864	1008	1152	1296	1440	1584	1728	1872	2016	2160	2304	2448	2592	2736	2880	3024	3168
4	72	144	216	288	360	432	504	576	648	720	792	864	936	1008	1080	1152	1224	1296	1368	1440	1512	1584
3	36	72	108	144	180	216	252	288	324	360	396	432	468	504	540	576	612	648	684	720	756	792
2	18	36	54	72	90	108	126	144	162	180	198	216	234	252	270	288	306	324	342	360	378	396
1	9	18	27	36	45	54	63	72	81	90	99	108	117	126	135	144	153	162	171	180	189	198
B	1	2	3	4	5	6	7	8	9	10	11	12	13	14	15	16	17	18	19	20	21	22

Notes:

- The first three white cell numbers in row # 3 have a sum of **612** which is the **apothem** length of the Great Pyramid measured in feet.   i.e. (144′ + 216′ + 252′) = **612** feet.

- The next four white cell numbers in row # 3, have a sum of **1512** which is twice the base length of the Pyramid, measured in feet, sometimes referred to as a quadrant. (324′ + 360′ + 396′ + 432′) = **1512** feet.

- The next three hi-lighted numbers in row # 3, 612 , 720 and 756 are the apothem length , the corner line length and the apothem length of the Great Pyramid i.e. **apothem = 612** feet., **corner-line = 720** feet, **baseline width = 756** feet. True compass bearings for the four faces of the Great Pyramid are highlighted in gray in row # 2. **East, South, West and North.** — 90 ° - 180° - 270° - 360° respectively.

TABLE -4

127

Table-4A — The Sequential Multiples of Nine and Application of the Trigonometric Phi (φ) Function Rule

	1	2	3	4	5	6	7	8	9	10	11	12	13	14	15	16	17	18	19	20	21
Sign RH6	cos 288	cos 576	cos 864	cos 1152	cos 1440	cos 1728	cos 2016	cos 2304	cos 2592	cos 2880	cos 3168	cos 3456	cos 3744	cos 4032	cos 4320	cos 4608	cos 4896	cos 5184	cos 5472	cos 5760	cos 6048
f(φ)	1/2φ	-φ/2	-φ/2	1/2φ	+1	1/2φ	-φ/2	-φ/2	1/2φ	+1	1/2φ	1/2φ	-φ/2	1/2φ	+1	1/2φ	-φ/2	-φ/2	1/2φ	+1	1/2φ
Rev	0.8	1.6	2.4	3.2	4.0	4.8	5.6	6.4	7.2	8.0	8.8	9.6	10.4	11.2	12.0	12.8	13.6	14.4	15.2	16.0	16.8
Sign RH5	cos 144	cos 288	cos 432	cos 576	cos 720	cos 864	cos 1008	cos 1152	cos 1296	cos 1440	cos 1584	cos 1728	cos 1872	cos 2016	cos 2160	cos 2304	cos 2448	cos 2592	cos 2736	cos 2880	cos 3024
f(φ)	-φ/2	1/2φ	1/2φ	-φ/2	+1	-φ/2	1/2φ	1/2φ	-φ/2	+1	-φ/2	1/2φ	1/2φ	-φ/2	+1	-φ/2	1/2φ	1/2φ	-φ/2	+1	-φ/2
Rev	0.4	0.8	1.2	1.6	2.0	2.4	2.8	3.2	3.6	4.0	4.4	4.8	5.2	5.6	6.0	6.4	6.8	7.2	7.6	8.0	8.4
Sign RH4	cos 72	cos 144	cos 216	cos 288	cos 360	cos 432	cos 504	cos 576	cos 648	cos 720	cos 792	cos 864	cos 936	cos 1008	cos 1080	cos 1152	cos 1224	cos 1296	cos 1368	cos 1440	cos 1512
f(φ)	1/2φ	-φ/2	-φ/2	1/2φ	+1	1/2φ	-φ/2	-φ/2	1/2φ	+1	1/2φ	-φ/2	-φ/2	1/2φ	+1	-φ/2	-φ/2	1/2φ	1/2φ	+1	1/2φ
Rev	0.2	0.4	0.6	0.8	1.0	1.2	1.4	1.6	1.8	2.0	2.2	2.4	2.6	2.8	3.0	3.2	3.4	3.6	3.8	4.0	4.2
Sign RH3	cos 36	cos 72	cos 108	cos 144	cos 180	cos 216	cos 252	cos 288	cos 324	cos 360	cos 396	cos 432	cos 468	cos 504	cos 540	cos 576	cos 612	cos 648	cos 684	cos 720	cos 756
f(φ)	φ/2	1/2φ	-1/2φ	-φ/2	-1	-φ/2	-1/2φ	1/2φ	φ/2	+1	φ/2	1/2φ	-1/2φ	φ/2	-1	-φ/2	-1/2φ	1/2φ	φ/2	1	φ/2
Rev	0.1	0.2	0.3	0.4	0.5	0.6	0.7	0.8	0.9	1.0	1.1	1.2	1.3	1.4	1.5	1.6	1.7	1.8	1.9	2.0	2.1
Sign RH2	sine 18	cos 36	sine 54	cos 72	sine 90	cos 108	sine 126	cos 144	sine 162	cos 180	sine 198	cos 216	sine 234	cos 252	sine 270	cos 288	sine 306	cos 324	sine 342	cos 360	sine 378
f(φ)	1/2φ	φ/2	φ/2	1/2φ	+1	-1/2φ	φ/2	-φ/2	1/2φ	-1	-1/2φ	-φ/2	-φ/2	-1/2φ	-1	1/2φ	φ/2	φ/2	-1/2φ	+1	1/2φ
Rev	0.05	0.1	0.15	0.2	0.25	0.3	0.35	0.4	0.45	0.5	0.55	0.6	0.65	0.7	0.75	0.8	0.85	0.9	0.95	1.0	1.05
Sign RH1	n/a 9	sine 18	n/a 27	cos 36	n/a 45	sine 54	n/a 63	cos 72	n/a 81	sine 90	n/a 99	cos 108	n/a 117	sine 126	n/a 135	cos 144	n/a 153	sine 162	n/a 171	cos 180	n/a 189
f(φ)	n/a	1/2φ	n/a	φ/2	n/a	φ/2	n/a	1/2φ	n/a	+1	n/a	-1/2φ	n/a	φ/2	n/a	-φ/2	n/a	1/2φ	n/a	-1	n/a
Rev	0.025	0.05	0.075	0.1	0.125	0.15	0.175	0.2	0.225	0.25	0.275	0.3	0.325	0.35	0.375	0.4	0.425	0.45	0.475	0.5	0.525
Base	1	2	3	4	5	6	7	8	9	10	11	12	13	14	15	16	17	18	19	20	21

Phi function identity rules

1. If (n) is an integer divisible by **9**, and (n) ÷**360** contains **one decimal place**, (i.e., **.1, .2, .3, .4, .6, .7, .8, .9**), excluding (.0 or .5), then the **Cosine** of (n) can be expressed as a function of Phi. - i.e. cos.(n) = f (φ). --- If (n) ÷360 ends with (.0 or .5), then cos.(n) = ± 1.

2. If (n) ÷**360** has **two decimal places** and is an **odd multiple** of (.05), (i.e., .05, .15, .35, .45, .55, .65, .85, .95), excluding (.25 & .75), then **Sine** of (n) can be expressed as a function of Phi. -- i.e. sin.(n) = f (φ). -- If (n) ÷360 ends with (.25 or .75), then **sin.(n) = ±1**.

3. Numbers (n) that end with a **4** or **6** have a trig function of ± φ/2. & Numbers (n) that end with a **2** or **8** have a trig function of ± 1/(2φ).

©2001, Joseph Turbeville

129

Alpha-Numeric Conversion
Distillation Order (1 – 9)

A 1 J 10	S 19	BB 28	KK 37	TT 46	
B 2	K 11 T 20	CC 29	LL 38	UU 47	**B·K·T** = 2·11·20 = **440** Base Width in cubits of the Great Pyramid at Giza
C 3	L 12	U 21 DD 30	MM39	VV 48	**C·L·U** = 3·12·21 = **756** Base Width in feet of the Great Pyramid at Giza
D 4	M 13	V 22	EE 31 NN 40	WW 49	
E 5	N 14	W 23	FF 32	OO 41 XX 50	
F 6	O 15	X 24	GG 33	PP 42	YY 51 60 — Diameter of Moon **F·O·X**=6·15·24 = **2160** in S. Miles
G 7	P 16	Y 25	HH 34	QQ43	ZZ 52 — **G·P·Y**=7·16·25 = **2800** Height-Great Pyramid **280** cubits (Base 10 multiple)
H 8	Q 17	Z 26	II 35	RR 44	
I 9	R 18	AA 27	JJ 36	SS 45	**I·R** = 9 ·18 = **162** ln (440 cu./**162** cu) = 0.9992 ≅ 1 ln (2.7182818280) = 1.0000

TABLE-5

130

131

ALPHA-NUMERIC VALUES (ANV) OF ENGLISH ALPHABET - TRIPLE LETTER SUMS, PRODUCTS & SQUARES - DISTILLED

Table-5B

ANV COLUMNS (1)	(2)	(3)	=	Sum	Prod.	Dist.	(1) Sq.	(2)Sq.	(3)Sq.	=	Sum	Prod.	Dist.	Squared Product Pairs
A 1	J 10	S 19	=	30	190	1	1	100	361	=	462	36,100	1	361 x (10)² / 100 x (19)²
B 2	K 11	T 20	=	33	440	8	4	121	400	=	525	193,600	1	400 x (22)² / 121 x (40)²
C 3	L 12	U 21	=	36	756	9	9	144	441	=	594	571,536	9	441 x (36)² / 144 x (63)²
D 4	M 13	V 22	=	39	1144	1	16	169	484	=	669	1,308,736	1	484 x (52)² / 169 x (88)²
E 5	N 14	W 23	=	42	1610	8	25	196	529	=	750	2,592,100	1	529 x (70)² / 196 x (115)²
F 6	O 15	X 24	=	45	2160	9	36	225	576	=	837	4,665,600	9	576 x (90)² / 225 x (144)²
G 7	P 16	Y 25	=	48	2800	1	49	256	625	=	930	7,840,000	1	625 x (112)² / 256 x (175)²
H 8	Q 17	Z 26	=	51	3536	8	64	289	676	=	1029	12,503,296	1	676 x (136)² / 289 x (208)²
I 9	R 18	? 27	=	54	4374	9	81	324	729	=	1134	19,131,876	9	324 x (243)² / 729 x (162)²

First 24 Fibonacci Numbers	Distilled Value
1 : 1 = 1	1
2: 1 = 1	1
3 : 2 = 2 Prime	2
4 : 3 = 3 Prime	3
5 : 5 = 5 Prime	5
6 : 8 = 2^3	8
7 : 13 = 13 Prime	4
8 : 21 = 3•7	3
9 : 34 = 2•17	7
10 : 55 = 5•11	1
11 : 89 = 89 Prime	8
12 : 144 = 2^4•3^2	9
13 : 233 = 233 Prime	8
14 : 377 = 13•29	8
15 : 610 = 2•5•61	7
16 : 987 = 3•7•47	6
17 : 1597 = 1597 Prime	4
18 : 2584 = 2^3•17•19	1
19 : 4181 = 37•113	5
20 : 6765 = 3•5•11•41	6
21 : 10946 = 2•13•421	2
22 : 17711 = 89•199	8
23 : 28657 = 28657 Prime	1
24 : 46368 = 2^5•3^2•7•23	9
Sum	117

TABLE-6

- If the Fibonacci series is continued, the distilled values of the numbers will repeat those shown above every 24 cycles.

216 Cells arranged with **9** Rows of the distilled **Products** of the original **24** distilled Fibonacci Numbers of Table-1.   Each successive row is obtained by multiplying the square of the row number by each respective top row factor and distilling.   Some cells in the Table have been darkened to aid in the discussion of historically significant numbers discovered in this Table.

ROW SUM ô

															ROW SUM
															126
															126
															216
															126
															126
															216
															126
															126
															216
51	57	63	51	57	45	51	57	51	63	57	51	45	57	81	**1404**
234			198			270			234			198		270	
702							702								

©2001, Joseph Turbeville

- **719 ft** = corner line length of GP --- **378 ft** = Half Base width of GP --- Shown in "Tee" formation above.
- **6 MI Rows** sum to 126 = **756** --- Base width GP = **756 ft.** --- 2.1 rev. x 360° = 756°   ∴ cos 756° = φ/2.
- **MI Row Nos. 4 & 5** both have two of the number **756** --- 4 x 756 ft. = **3024 ft.** = Baseline perimeter of GP.
- **2 MI Rows** sum to 216 = **432** --- **43200 : 1** = Earth : GP scale --- 1.2 rev. x 360° = 432°   ∴ cos 432° = 1/(2 φ)
- Sum of **126 + 486 = 612** ft. = Apothem of GP. Also sqrt.(719² - 378²) = 611.6 ≅ **612** ft. = Apothem of GP.
- **198,234,270** seconds/ 86400 sec per day/ 365 day per year ≅ 2 Pyramid Pi years = **44/7** years = 6.285714286 year.

TABLE-7   DISTILLED ROW-DIGIT PRODUCTS OF PARENT TABLE-1